Elements of the Geometry and Topology of Minimal Surfaces in Three-Dimensional Space

Translations of

MATHEMATICAL MONOGRAPHS

Volume 93

Elements of the Geometry and Topology of Minimal Surfaces in Three-Dimensional Space

A. T. Fomenko
A. A. Tuzhilin

American Mathematical Society
Providence, Rhode Island

ЭЛЕМЕНТЫ ГЕОМЕТРИИ И ТОПОЛОГИИ
МИНИМАЛЬНЫХ ПОВЕРХНОСТЕЙ
В ТРЕХМЕРНОМ ПРОСТРАНСТВЕ

А. Т. Фоменко и А. А. Тужилин

Translated by E. J. F. Primrose from an original Russian manuscript
Translation edited by Simeon Ivanov

2000 *Mathematics Subject Classification.* Primary 53Axx;
Secondary 49–XX, 57–XX.

ABSTRACT. This book arose as a result of writing up the lectures of A. T. Fomenko delivered at the Moscow Mathematical Society (in Moscow University) for students of mathematics, physics and mechanics (in the framework of the cycle of lectures "Readings for students"). The writing up of the lectures for the press has been carried out by A. A. Tuzhilin and A. T. Fomenko.

For additional information and updates on this book, visit
www.ams.org/bookpages/mmono-93

Library of Congress Cataloging-in-Publication Data

Fomenko, A. T.

[Ėlementy geometrii i topologii minimal′nykh poverkhnosteĭ v trekhmernom prostranstve. English]

Elements of the geometry and topology of minimal surfaces in three-dimensional space / A. T. Fomenko and A. A. Tuzhilin.

p. cm. — (Translations of mathematical monographs; v. 93)

Translation of: Ėlementy geometrii i topologii minimal′nykh poverkhnosteĭ v trekhmernom prostranstve.

Includes bibliographical references and indexes.

ISBN 0-8218-4552-7 (acid free)

ISBN 0-8218-3791-5 (softcover)

1. Surfaces, Minimal. I. Tuzhilin, A. A. II. Title. III. Series.

QA644.F6513 1991 91-21604
516.3′62—dc20 CIP

---

10 9 8 7 6 5 4 3 2 10 09 08 07 06 05

Contents

Introduction

This book arose as a result of writing up the lectures of A. T. Fomenko delivered at the Moscow Mathematical Society (in Moscow University) for students of mathematics, physics and mechanics (in the framework of the cycle of lectures "Readings for students"). The writing up of the lectures for the press has been carried out by A. A. Tuzhilin and A. T. Fomenko.

The reader who wishes to extend his study of the theory of minimal surfaces can turn to the following more special books and articles:

H. Federer, *Geometric measure theory*, Springer-Verlag, New York, 1969.

J. C. C. Nitsche, *Vorlesugen über Minimalflächen*, Springer-Verlag, Berlin-Heidelberg-New York, 1975.

E. Giusti, *Minimal surfaces and functions of bounded variation*, Birkhäuser, Boston-Basel-Stuttgart, 1984.

A. T. Fomenko, *Topological problems in topology. The geometry of length, area and volume*, Moscow University Press, Moscow, 1984; English transl., Gordon and Breach, 1990.

A. T. Fomenko, *Variational principles in topology*, "Nauka", Moscow, 1982; English transl., *Multidimensional minimal surface theory*, Kluwer, 1990.

Dào Trong Thi and A. T. Fomenko, *Minimal surfaces and Plateau's problem*, "Nauka", Moscow, 1987; English transl., Amer. Math. Soc., Providence, RI, 1991.

Yu. A. Aminov, *Minimal surfaces*, Kharkov University Press, Kharkov, 1978. (Russian)

R. Finn, *Equilibrium capillary surfaces*, Grundlehren der Mathematischen Wissenschaften 284, Springer-Verlag, New York, 1986.

A. T. Fomenko, *The Plateau problem*, 2 vols., Gordon and Breach, London, 1990.

R. Osserman, *A survey of minimal surfaces*, Van Nostrand Reinhold, New York-London-Melbourne, 1969.

R. Courant, *Dirichlet's principle, conformal mapping, and minimal surfaces*, Interscience, New York, 1950.

CHAPTER 1

Physical Prerequisites

§1. Interfaces between two media

The theory of minimal surfaces and surfaces of constant mean curvature is a branch of mathematics that has been intensively developed, particularly recently. On the basis of this theory we can investigate soap films and soap bubbles, interfaces between two media, which occur widely in chemistry and biology, for example, membranes in living cells, capillary phenomena, and so on. Minimal surfaces also turn out to be useful in architecture. Before giving exact definitions, let us consider some examples.

1. Soap films and soap bubbles. If we dip a wire contour into soapy water, and then carefully lift it out, a soap is left on the contour. For many "young researchers" this is the time to obtain an iridescent bubble by blowing on the film. However, the soap films themselves conceal unexpected properties, which we can immediately see by making a simple experiment. Bend a wire contour, as shown in Figure 1. This is a so-called Douglas contour. Let u and v denote the distances between the extreme circles of the contour. It turns out that by lifting the contour out of the soapy water differently we can obtain different soap films. Figure 2 shows some of them (for small $u = u_0$ and $v = v_0$).

If the contour is not deformed in the process of obtaining the film, then as a rule films of type 1a and 1b are formed. A film of type 2 is obtained if at the time of lifting the contour out of the soapy water the left and right circles are kept joined $(u = 0, v = 0)$, but after lifting the contour out they are let free. The elastic contour returns to its original position $(u = u_0, v = v_0)$, and a film of type 2 remains on it. Films of type 3a and 3b can be obtained similarly, by combining only right (or only left) circles. To obtain a film of

FIGURE 1

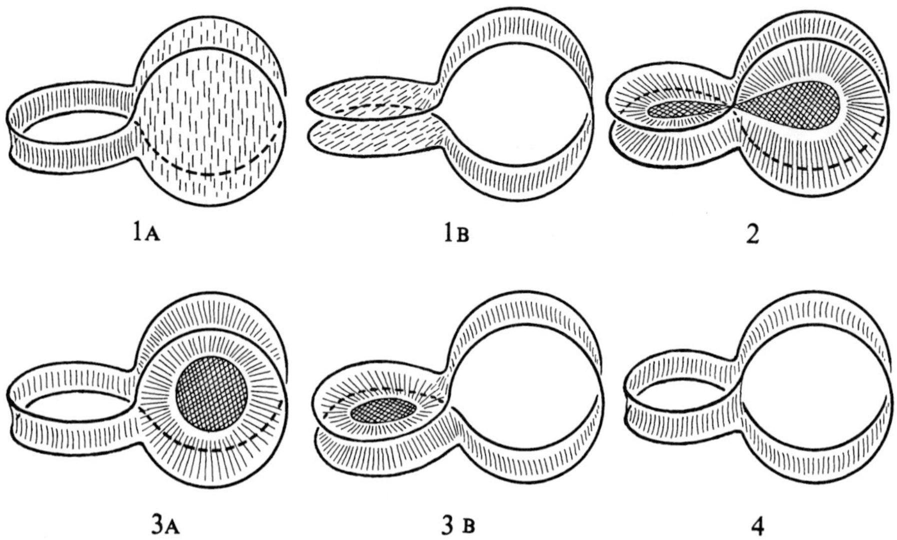

FIGURE 2

type 4 it is sufficient to puncture a film of type 3a by a disc D^2.

We observe that films of type 2 and 3 are not smooth surfaces. They have singularities: a singular point A where four singular edges meet (type 2), or a singular edge S^1 (type 3). Moreover, smooth films of types 1 and 4 have different topological type: a film of type 1 is a two-dimensional disk D^2, while a film of type 4 is a torus with a point deleted (see Figure 3).

Thus, on a given contour we can, generally speaking, stretch many soap

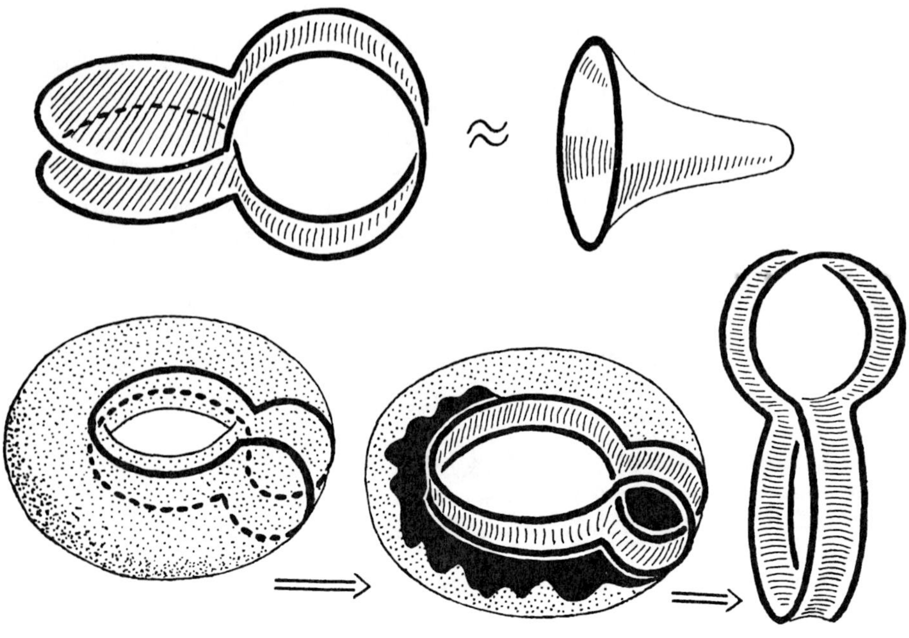

FIGURE 3

films, and not every film need be a smooth surface. Moreover, on a contour that is a bent circle we can stretch a minimal torus with a point deleted (a film of type 4). How many soap films can span a given contour? What topological types of films can occur? What singularities can be found on a soap film? These and other questions are considered in the theory of minimal surfaces (see below).

Soap bubbles and soapsuds, that is, a system of soap bubbles, are just as interesting to the researcher. Physical bubbles differ from soap films in that bubbles bound regions of space in which air is under greater pressure than the air outside. If we look closely at the soapsuds, we observe that the singularities at the junction of different bubbles are similar to the singularities of soap films. By carrying out many experiments with the interfaces bewteen two media the Belgian physicist Joseph Plateau (1801–1883) formulated four principles, which describe the possible stable singularities on these surfaces. It turns out that the two types of singularities revealed above—a smooth singular edge at which three sheets meet, and a singular point at which four smooth singular edges meet, each pair of which is spanned by a smooth sheet—are the only possible singularities of stable soap films and soapsuds (Plateau's 1st, 2nd, and 3rd principles). Moreover, in the first case the sheets meet at a singular edge at an angle of 120°, while in the second case the singular edges meet at a point at an angle of about $109^\circ\, 28'\, 16''$ (more precisely, the cosine of this angle is $-1/3$), like the four line segments drawn from the center of a regular tetrahedron to its vertices (Plateau's 4th principle). We give some elementary justification of these principles below, first describing the variational principle that underlies soap films and soap bubbles.

2. The Poisson-Laplace theorem. Soap films and soap bubbles can be regarded as the interfaces between two homogeneous media in equilibrium. A soap film with boundary locally, that is, in a neighborhood of each of its points, separates two media, air-air, in each of which the pressure is the same. Therefore the total pressure on each small area of a soap film is zero. In a soap bubble the pressure inside is greater than the pressure outside, so the vector of the resultant pressure is directed outwards. This force must be compensated by the forces of surface tension. Since the pressure is always directed along the normal to the interface and is the same in absolute value at all points of this interface because of the homogeneity of the media, the interface is "curved in the mean" identically at all its points. To give a precise meaning to this statement, we need to define the geometric concept of "mean curvature" (for the details see [1], [15]).

Let M be a smooth two-dimensional surface in $\mathbb{R}^3$, suppose that the point P lies on the surface M, and that $N(P)$ is one of the two unit normals to M at P (the vector $N(P)$ is orthogonal to the tangent plane to M passing through the point P; we denote this tangent plane by T_PM).

Through P we pass a plane Π containing to $N(P)$. The plane Π intersects M along a curve γ called a *normal section*. The unit vector v tangent

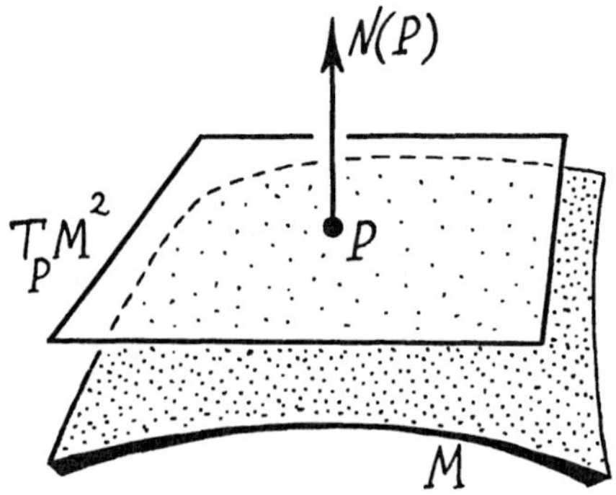

FIGURE 4A

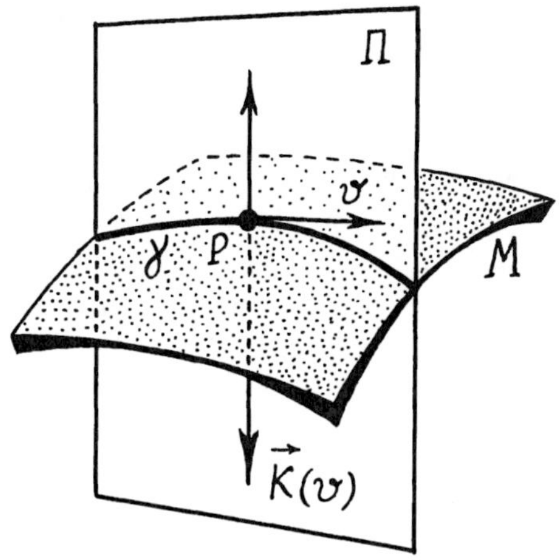

FIGURE 4B

to γ at P is called a direction of this normal section. Clearly, the vector $-v$ determines the opposite direction of the normal section γ, and the vectors v and $-v$ lie in the tangent plane T_PM (Figure 4).

Let $\vec{\kappa}(v)$ denote the curvature vector of γ in the direction v, that is, the acceleration vector at P under motion along γ with unit speed. We note that $\vec{\kappa}(v) = \vec{\kappa}(-v)$. It is easy to show that the curvature vector $\vec{\kappa}(v)$ is collinear with the normal $N(P)$. We define the curvature $\kappa(v)$ of a normal section γ in the direction v with respect to the normal $N(P)$ as the quantity $\kappa(v) = \langle \vec{\kappa}(v), N(P) \rangle$, where the brackets $\langle\ ,\ \rangle$ denote the standard scalar product of vectors in $\mathbb{R}^3$. Clearly, the continuous function $\kappa(v)$ takes maximum and minimum values (since $\kappa(v)$ is a function on the (compact) circle S^1 formed by all directions v). These values are called the *principal curvatures* κ_1 and κ_2 of the surface M at the point P, and the normal sections in which the values κ_1 and κ_2 are attained are called the *principal sections.*

DEFINITION. The *mean curvature* H of a surface M at a point $P \in M$ with respect to the normal $N(P)$ is half the sum of the principal curvatures: $H = (\kappa_1 + \kappa_2)/2$.

Let φ denote the angle between the direction v of an arbitrary normal section γ at a point $P \in M$ and the direction of the principal section γ_1 (Figure 5).

If κ_1 is the curvature of the principal section γ_1, and κ_2 is the other principal curvature, then by Euler's well-known formula

$$\kappa(v) = \kappa(\varphi) = \kappa_1 \cos^2 \varphi + \kappa_2 \sin^2 \varphi .$$

To represent more clearly the distribution of the curvatures of the normal sections γ as the angle φ changes, we construct in the plane with polar coordinates (ρ, φ) the graph $\rho = |\kappa(\varphi)|$. We can distinguish the following

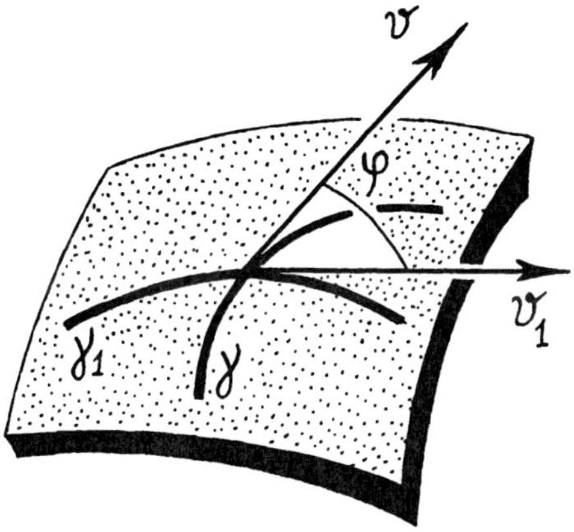

FIGURE 5

cases:

a) κ_1 and κ_2 are nonzero and have the same sign. In this case the graph is an ellipse with semiaxes $|\kappa_1|$ and $|\kappa_2|$. If $\kappa_1 = \kappa_2 \neq 0$, the ellipse degenerates to a circle (Figure 6a);

b) κ_1 and κ_2 are nonzero and of different signs. In this case the graph is a "four-leafed rose" (Figure 6b);

c) one of the curvatures κ_i is zero—the "four-leafed rose" degenerates into a "two-leafed rose" (Figure 6c);

d) both the principal curvatures are zero. The graph is a point (the origin).

It is now clear that if κ_1 is not equal to κ_2, then there are exactly two principal sections, which are in fact orthogonal to each other. If the principal curvatures are equal, then the curvature of all the normal sections is the same and equal to the mean curvature H.

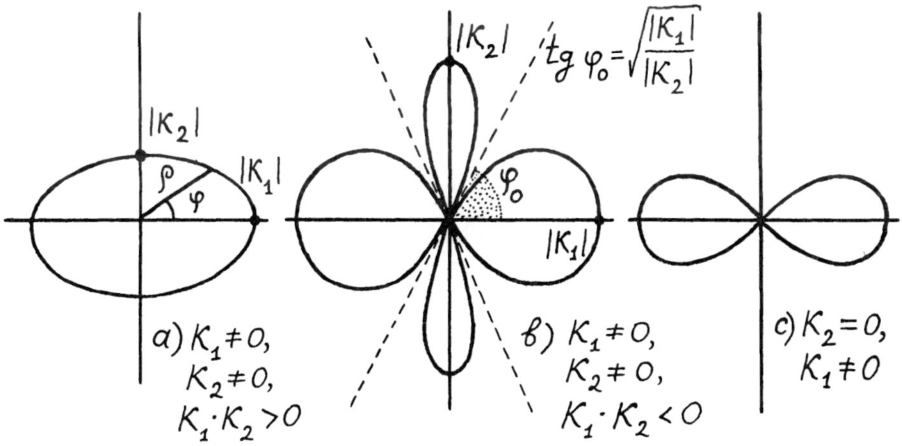

FIGURE 6

Exercise. Prove that half the sum of the curvatures of any two mutually orthogonal normal sections is constant and equal to H.

The Poisson-Laplace theorem. *Suppose that a smooth two-dimensional surface M in $\mathbb{R}^3$ is the interface between two homogeneous media in equilibrium. Let P_1 and P_2 be the pressures in the media. Then the mean curvature H of the surface M is constant and equal to $H = h(P_1 - P_2)$, where the constant $\lambda = 1/h$ is called the coefficient of surface tension, and $P_1 - P_2$ is the difference between the pressures in the media* (*the resultant pressure*).

Thus, the expression "curved in the mean identically" implies that the mean curvature of the surface is constant. Taking account of what we said above, we can conclude that the mean curvature H of a soap film is zero, $H \equiv 0$, and the mean curvature H of a soap bubble is a constant $\neq 0$. In mathematics surfaces with $H = \text{const}$ are called surfaces of constant mean curvature. For the case $H = 0$ these surfaces have a special name—minimal surfaces (the reason for the origin of the name is explained in §2). Sometimes they are called soap films, and surfaces with $H = \text{const} \neq 0$ are called soap bubbles.

Surfaces of constant mean curvature are widespread in nature and play an important role in various research. Thus, for example, surface interactions on the interface between two media determine the character and rate of chemical reactions. Various membranes, such as the ear-drum and membranes that separate living cells, are minimal surfaces. One more example consists of microscopic marine animals—Radiolaria (see [2]).

§2. The principle of economy in Nature

In this section we talk about an alternative approach to the description of minimal surfaces and surfaces of constant mean curvature, based on the variational principle (for a more detailed historical survey see [3]).

1. Optimality and Nature. In 1744 the French scientist Pierre-Louis-Moreau de Maupertuis put forward his famous principle, which has become known as the principle of least action. In 1746 Maupertuis published a paper "The laws of motion and rest deduced from a metaphysical principle". This metaphysical principle is based on the assumption that Nature always acts with the greatest economy. Starting from this position, Maupertuis draws the following conclusion: if certain changes occur in Nature, then the total action needed to carry out these changes must be as small as possible.

In parallel and independently Leonhard Euler in 1744 obtained a strict proof that the principle of least action can be used to describe the motion of a material point in a field of conservative forces such as the motion of the planets around the Sun. Euler also put forward the conjecture that for any phenomenon in the Universe we can find a maximum or minimum rule to which it is subject. This remark appeared in the Appendix to his famous work of 1743 "Methods of finding curves that are subject to a maximum or minimum property", the first textbook on the calculus of variations.

When in 1746 Maupertuis published his work on the principle of least action he was well aware of Euler's achievement, since he briefly described it in the Preface. Then, however, he added: "This remark ... is a beautiful application of my principle to the motion of the planets", thus asserting his priority.

Euler reacted to this remark by giving up his right of priority, for which he was strongly criticized by certain historians of science. We shall not go into the details of the subsequent keen discussion that developed over priority in the discovery of the principle of least action. We shall only say that other people (König and apparently Leibniz) laid claim to authority and that the discussion was linked to the anthropomorphic understanding of the terms "living force" and "action" and with theology (see [3], [37]).

However, we observe that Maupertuis, who formulated his principle starting from the idea of the perfection of God, tested it on a few examples, but he did not investigate some of them thoroughly. It turns out that the principle of least action is not always true.

Let us consider one of the examples given by Maupertuis—the reflection of light. Here the law of least action leads to the conclusion that a ray of light "selects" from all possible routes from the source to the receiver the one that can be covered in the least time (this rule had already been formulated by Fermat). If light is propagated in a homogeneous medium, then this minimum principle leads to the simpler rule: a ray of light moves along the shortest path joining the source and the receiver.

Consider a spherical mirror situated in a homogeneous medium. Suppose that the source S and the receiver T are symmetrical about some line l passing through the center of the sphere. What characterizes the trajectory SMT of the motion of light emitted from the source S, reflected in the mirror at a point M, and received by the receiver T? In this situation we apply the famous law presented in the work Catoptrica attributed to Euclid—the so-called law of reflection. The following brief formulation of it is well known: the angle of the incidence is equal to the angle of reflection.

Figure 7 shows two situations: a convex mirror (Figure 7a) and a concave mirror (Figure 7b). In both cases the point M is the point of intersection of the mirror and the line l (we consider the trajectories of the motion of light in the plane passing thorugh the source, the receiver, and the line l). Through M we draw the ellipse with foci at S and T. If M_1 is an arbitrary point lying outside the ellipse, M_2 is an arbitrary point inside it, and M is an arbitrary point on it, then $|M_1S|+|M_1T| > |MS|+|MT| > |M_2S|+|M_2T|$ (verify this).

Hence it follows that in the case of a convex mirror the trajectory SMT actually has the shortest length, but for a concave mirror this is not always

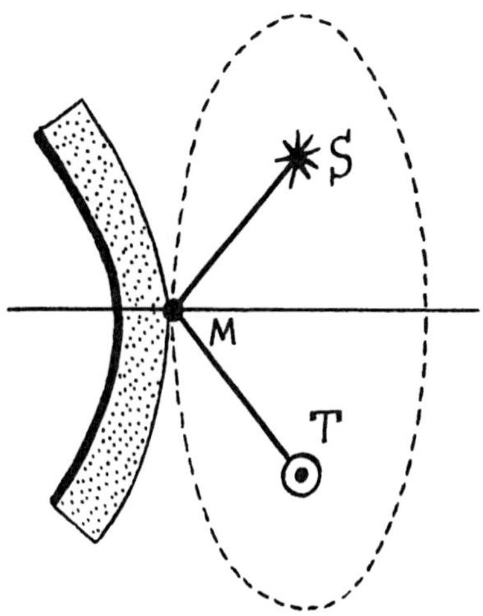

FIGURE 7A

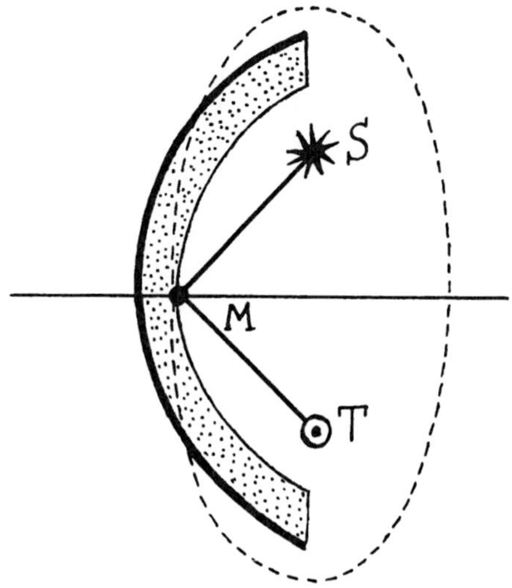

FIGURE 7B

so. In Figure 7b the source S and the receiver T are symmetrical about the center of the sphere. It is easy to see that any path SM_2T, where $M_2 \neq M$, is shorter than the path SMT.

Thus, whether Nature is most economical or most wasteful depends on the form of the mirror. Citing similar arguments, d'Arcy showed in 1749 and 1752 that the principle of Maupertuis was not clearly formulated, and leads to incorrect assertions.

Nevertheless, the idea of optimality of the phenomena of Nature plays an important role in physics. The mathematical formulation of this idea has given birth to the calculus of variations, the founders of which are usually taken to be Lagrange and three Swiss mathematicians from Basel: the brothers Johann and Jakob Bernoulli and a student of Johann Bernoulli—Leonhard Euler.

2. Minimal surfaces and optimality. It turns out that the forms of soap films are also optimal in a certain sense, namely the corresonding minimal surfaces are the extremals of the area functional. Let us explain this statement. For this we consider a soap film covering a given contour. Surface tension leads to the film tending to take up a form with the least possible surface energy (of course, this is an approximation, which nevertheless works well in practice). Since the surface energy is directly proportional to the surface area, as a result of minimizing the surface energy the area of the soap film is least compared to the areas of all sufficiently neighboring surfaces covering the given contour.

Thus, soap films are local minima of the area functional. However, minimal surfaces, that is, surfaces of zero mean curvature, need not minimize the area even among all neighboring surfaces (with given boundary). To explain this we note that the concept of "neighboring surfaces" can be defined in different ways. We shall understand by a neighboring surface of a given surface M one obtained by a small deformation of M, leaving the boundary ∂M fixed. For a zero-dimensional surface M, that is, when M is a point, a small deformation is a small displacement of M. When the dimension of M is at least one, we can define two types of small deformations: a deformation with small amplitude, that is, when each point of M is displaced not far from its original position, and a deformation with small support, when only a sufficiently small region of M undergoes a deformation; the closure of such a deformed region is called the support of the deformation. A deformation with sufficiently small support always increases the area of the minimal surface, so for such deformations the minimal surfaces do minimize the area. For deformations with small amplitude this is not so. Nevertheless, for such deformations minimal surfaces are extremals (critical points) of the area functional.

It is easy to show that the (finite) area of any surface in $\mathbb{R}^3$ can always be increased by an arbitrarily small deformation of this surface, so no surface can be a local maximum for the area functional. It is well known that critical points other than a local minimum and a local maximum are called saddle points. Minimal surfaces corresponding to saddle points of the area functional (for deformations with small amplitude) are said to be unstable. If we have succeeded in creating a soap film that has the form of an unstable minimal surface, then fluctuations of this film that are small in amplitude, which always exist in the real world, would instantaneously lead to its collapse—the film constructed would turn out to be unstable.

Thus, minimal surfaces are critical points of the area functional. It turns out that the converse is true: a surface M that is a critical point of the area functional (considered on the space of all possible surfaces close (in amplitude) to M and having the same boundary ∂M) is minimal, that is, it has zero mean curvature.

Surfaces of constant mean curvature are also extremals of a certain functional. They can also be obtained as extremals of the area functional if we restrict the possible deformations. As an example we consider a soap bubble. If we blow on it, the film, sagging in one place, will swell at another place in such a way that the volume of the region inside the bubble is unchanged. This observation is the basis of the definition of a surface of constant mean curvature from the viewpoint of the variational principle. For a closed surface bounding a region in $\mathbb{R}^3$, as admissible deformations we consider only those that preserve the volume of the region bounded by this surface. The condition that the volume is preserved can be described in yet another way. For this we define the function of change of volume of a region V bounded by a surface M under a deformation M_t of this surface. For each $t = t_0$ we consider the totality of regions included between the surfaces M and M_{t_0}. From the total volume of those that lie outside V we subtract the total volume of those that lie inside V, and we call the resulting number the change in volume at the instant $t = t_0$. Varying t_0, we obtain a function which is called the change of volume function. Clearly, the regions lying inside and outside V are on opposite sides of M, and so they can be defined without the use of V.

This observation enables us to define the change of volume function for a deformation M_t of an unclosed surface M (such as the soap film bounded by a wire contour or a hemisphere having the equator as its boundary), but in this case we must require that the deformation M_t is fixed ($= 0$) on the boundary ∂M of the surface M.

We say that a deformation M_t of a surface M (fixed on ∂M if $\partial M \neq \varnothing$) preserves the volume if the change of volume function constructed from this deformation is identically zero. It turns out that *surfaces of constant mean curvature are critical points of the area functional restricted to the space of deformations that preserve the volume.*

Since surfaces of nonzero constant mean curvature are not minimal surfaces, they are also not critical points of the area functional considered on the space of all possible deformations (fixed on the boundary), not only those that preserve the volume. Thus, restriction of the space of admissible deformations naturally leads to an increase in the number of surfaces that are critical points of the area functional considered on this space.

A description of minimal surfaces as extremals of the area functional proves very useful. For example, this approach has given the possibility of solving the so-called Plateau problem, which consists in the following: for any contour, among all surfaces of given topological type that bound it, is there a surface of least area? A positive answer to this problem in the case when the contour is a simple rectifiable Jordan curve (it has finite length) and the surface has the topological type of the disk D^2 was obtained in 1928 by the young American mathematician Jesse Douglas. However, his proof turned out to be incomplete, and up to 1931 his paper had not appeared in print. At about the same time a solution of Plateau's problem, obtained by the Hungarian mathematician Tibor Radó, was published. In the following decades Jesse Douglas also solved a number of other problems that arose in the theory of minimal surfaces. In particular, the powerful mathematical technique that he developed enabled him to prove the existence of minimal surfaces of high genus spanning one or finitely many contours. For his achievement Dougals was awarded in 1936 the highest prize in mathematics—the Fields Medal.

In conclusion we give the one-dimensional version of Plateau's problem, the so-called Steiner problem, and show how from its solution there follows a proof of Plateau's empirical principles, which describe all possible singularities of soap films of stable minimal surfaces.

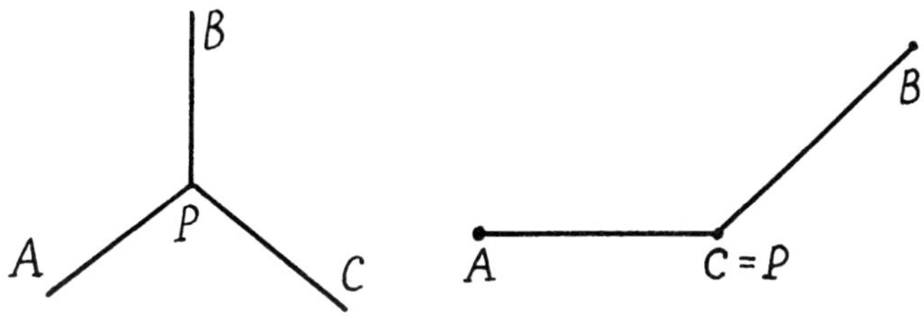

FIGURE 8

3. Steiner's problem. Let us begin with the simplest case. Suppose we need to connect the towns A, B, and C by a system of roads, that there are no obstructions, and that we are free to construct the roads where we like. Suppose that the region in which the towns lie is flat. The problem is to find a system of roads of least length. In mathematical language this means the following: for three given points A, B, and C lying in a plane, to find a point P and paths joining P to A, B, and C so that the total length of these paths shall be as small as possible. Since a line segment is the shortest path between its ends, the required paths are line segments PA, PB, and PC. It remains to choose P in an optimal way.

It turns out that the solution depends on the relative positions of A, B, and C. If all the interior angles of the triangle ABC are less than $120°$, then the required point P is uniquely determined from the condition that the angles $\widehat{APB}$, $\widehat{BPC}$, and $\widehat{CPA}$ are equal (they are thus equal to $120°$). But if one of the angles of the triangle ABC, say the angle at C, is at least $120°$, then P coincides with C (see Figure 8).

The proof of this assertion is based on some elementary geometrical lemmas.

LEMMA 1 (Heron's theorem). *Suppose that points A and B do not lie on a line a. Then among all the points P of the line a the point $P = P_0$ such that $|AP| + |BP|$ is as small as possible is uniquely determined from the following condition: the angle between AP_0 and the line a is equal to the angle between BP_0 and the line a (Figure 9).*

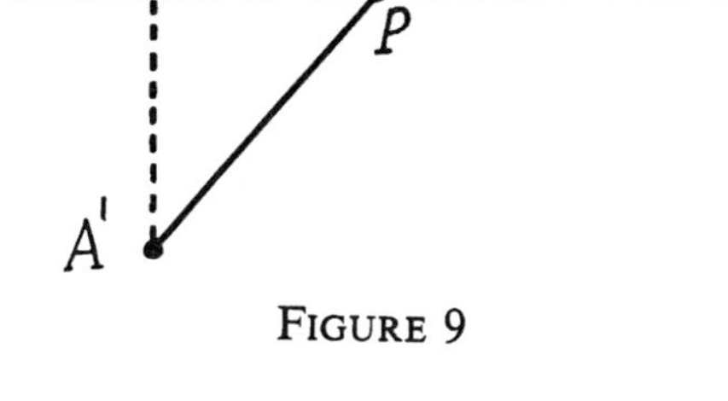

FIGURE 9

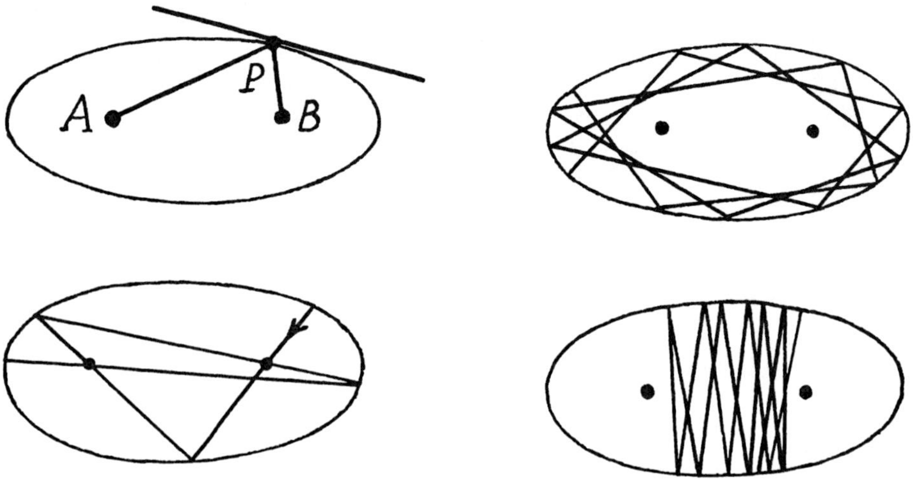

FIGURE 10

If A is a source of light, B is a receiver, and a is a mirror, then Heron's law can be regarded as a special case of the law of reflection (see above).

LEMMA 2. *Suppose that the points A and B are the foci of an ellipse, and that the line a touches this ellipse at a point P. Then the angles between the line a and the segments AP and BP are equal* (*Figure* 10).

It follows that if we put a source of light at one focus of an elliptical mirror, then all the rays collect at the other focus. Moreover, in elliptical billiards the ball always goes either outside the foci, or through the foci, or between the foci.

To prove Lemma 2 it is sufficient to observe that the sum of the distances from any point outside an ellipse to its foci is greater than the sum $|PA|+|PB|$ (since P lies on the ellipse).

Now suppose that P is a solution of Steiner's problem for a triangle ABC in which all the interior angles are less than 120°. Through P we draw the ellipse whose foci are A and B. Through P we draw the tangent a to this ellipse (see Figure 11). It is easy to see that CP is perpendicular to a. Therefore from Lemma 2 we see that $\widehat{APC} = \widehat{BPC}$. Similarly, $\widehat{BPC} = \widehat{APB}$. Thus, all the angles at P are equal to each other, and therefore equal to 120°. Now it is easy to construct the required point P and to see that the solution is unique.

If instead of three points A, B, and C we take any finite number of points, we obtain the generalized Steiner problem: it is required to join all

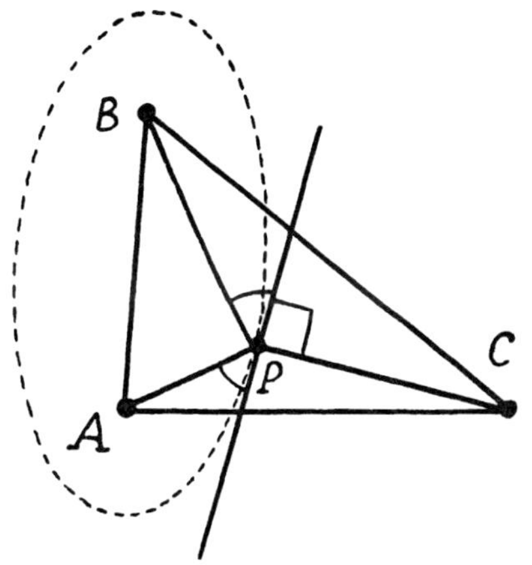

FIGURE 11

these points by a finite system of curves of least length. This problem can be restated as follows: how do we join n towns by a network of roads with the least expense? From the combinatorial point of view the solution of this problem is a combination of two solutions, obtained above for the case $n = 3$. Here are some examples (Figure 12).

We observe that the solution of the generalized Steiner problem is not unique. For example, if four points are vertices of a square, we can obtain two symmetrical solutions (Figure 13). We note that a system of paths of

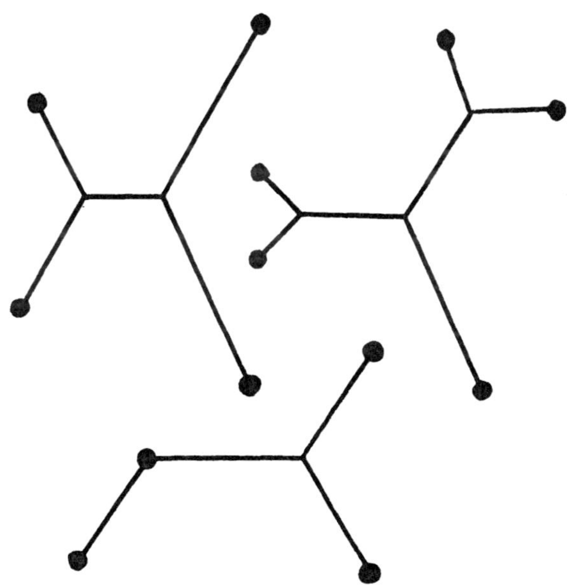

FIGURE 12

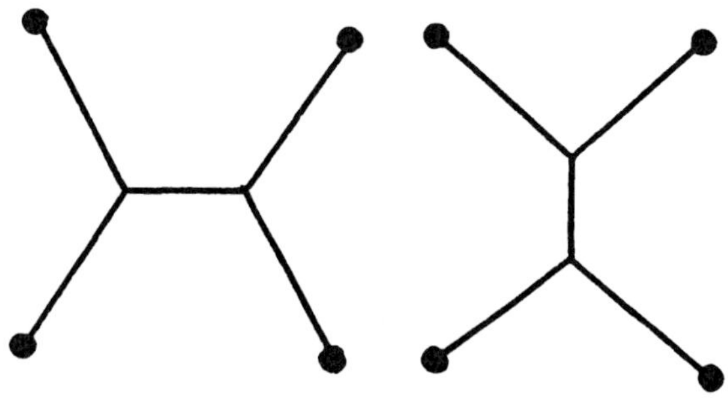

FIGURE 13

least length joining n points of a plane is called a *minimal network* (in the plane).

The generalized Steiner problem in the plane has still not been completely solved, so an experimental "solution" is of interest. Take two glass or transparent plastic sheets, place them in parallel planes, and join them by pieces of wire of the same length, equal to the distance between the sheets. Clearly, all the pieces of wire are parallel to one another and perpendicular to the sheets.

If we dip this configuration into soapy water, and carefully lift it out, then between the sheets there will be a soap film whose boundary consists of two parts: the set of joining wires and the set of "traces" which the film leaves on the sheets. We note that in accordance with the minimum principle the film is at an angle of 90° to each sheet. Moreover, this film consists of perpendicular sheets of planar rectangles that adjoin each other along the singular edges (see Figure 14). If a singular edge is not a joining wire, then in accordance with Plateau's principles exactly three rectangles meet on it at angles of 120°. We observe that the area of the resulting soap film is equal to the total length of the path joining the points where the wires are fastened (the "trace" of the soap film on a sheet) multiplied by the distance between the sheets. If the film has least area among all films with a partially free boundary consisting of the joining wires (rigid boundary) and the "traces" on the sheets (the hypothesis of the existence of such a film is called Plateau's problem with obstructions), then the corresponding "trace" on a sheet is a solution of the generalized Steiner problem for the configuration given by the fastening points.

Let us now turn to Plateau's principles, which describe all possible singularities of stable minimal surfaces. On a soap film we choose an arbitrary point P. We take a smaller and smaller neighborhood of P and blow it up to the same size. In the limit all the surfaces that join at P become planar, and the singular edges become segments of straight lines. Clearly, after such an enlargement the resulting fragment of stable film will also be stable. Now consider a sphere S^2 with center at P. Its intersection with the planar configuration we have constructed is a networks on the sphere S^2. Clearly

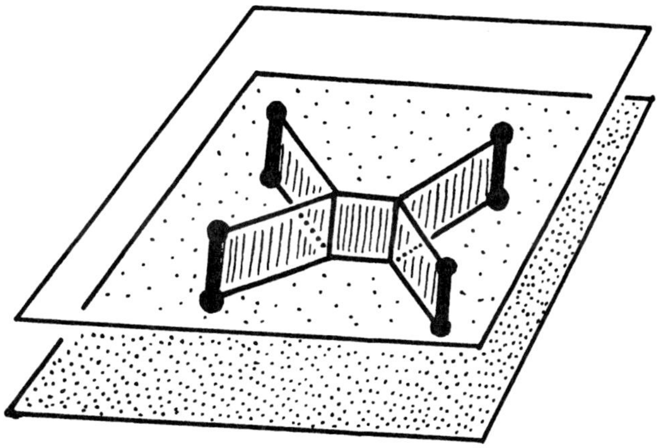

FIGURE 14

the curves of intersection are parts of great circles of S^2. Moreover, if l denotes the length of the network, and r is the radius of the sphere S^2, then the area s of the part of the planar configuration inside the sphere is equal to $s = lr/2$. Therefore from the stability of the film it follows that at each node of the network only three arcs can meet at angles of $120°$ (otherwise by a small deformation we could lower the length of the network, and hence the area of the film).

The following natural problem arises (the so-called Steiner problem on the sphere): to describe all possible networks on the sphere consisting of arcs of great circles meeting at each vertex of the network three at a time at equal angles (of $120°$). In contrast to the planar Steiner problem, the spherical problem has been completely solved. It turns out that there are exactly ten such networks, drawn in Figure 15.

A more careful analysis shows that only three of these ten networks (the first three in Figure 15) correspond to configurations that minimize the area. Figure 16 shows soap films stretched on contours corresponding to minimal networks on a sphere. Only the first three of them are cones corresponding to the local arrangement of soap films described in Plateau's principles. This observation is a physical "proof" that only in the first three cases are the cones absolutely minimal, that is, they correspond to singularities occurring in stable soap films.

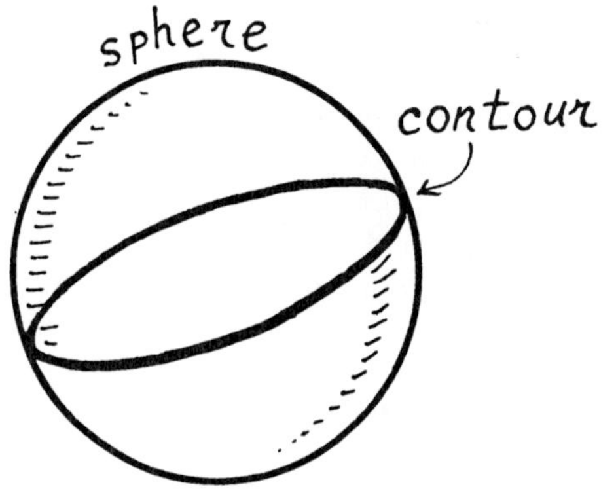

FIGURE 15A

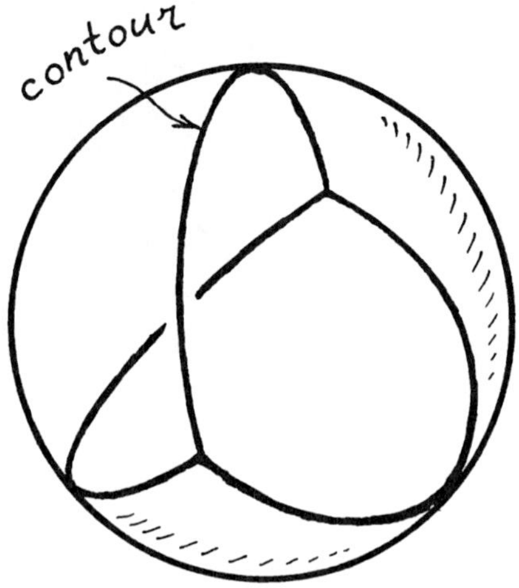

FIGURE 15B

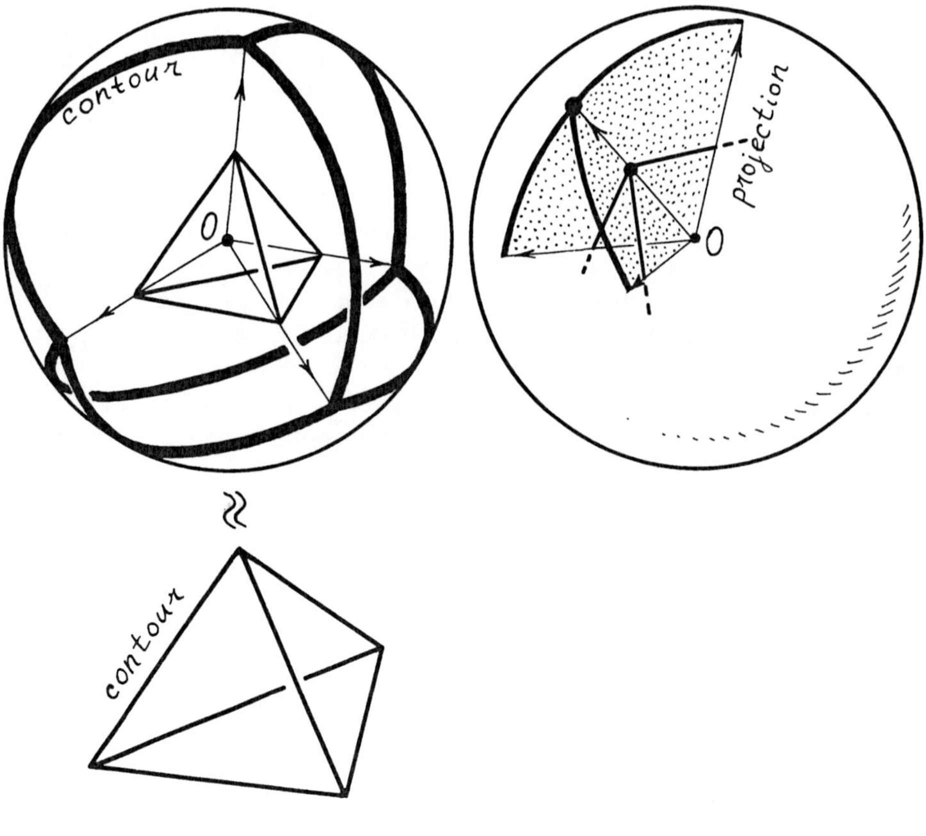

FIGURE 15C

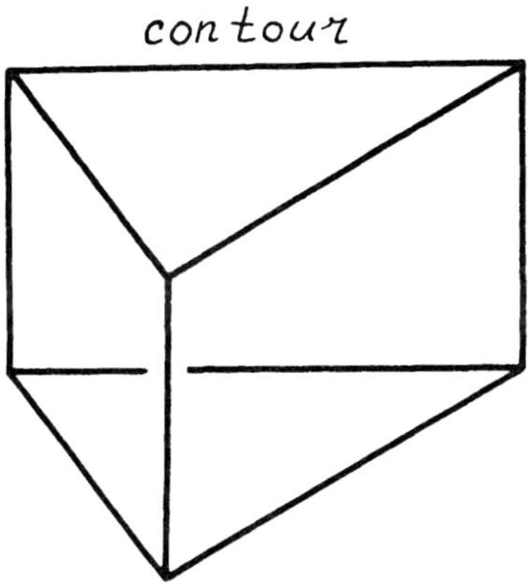

FIGURE 15D

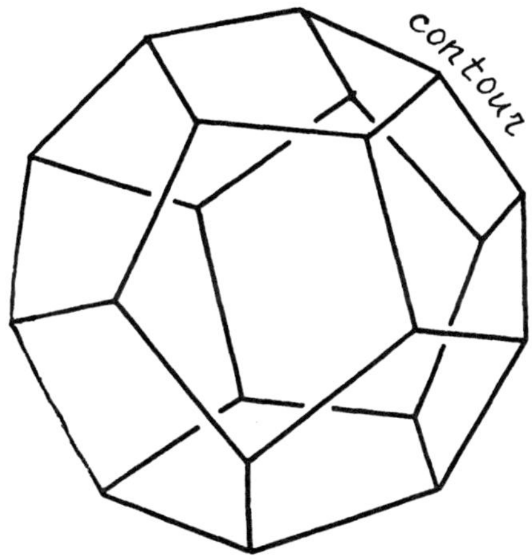

(minimal film ≠ cone)

FIGURE 15E

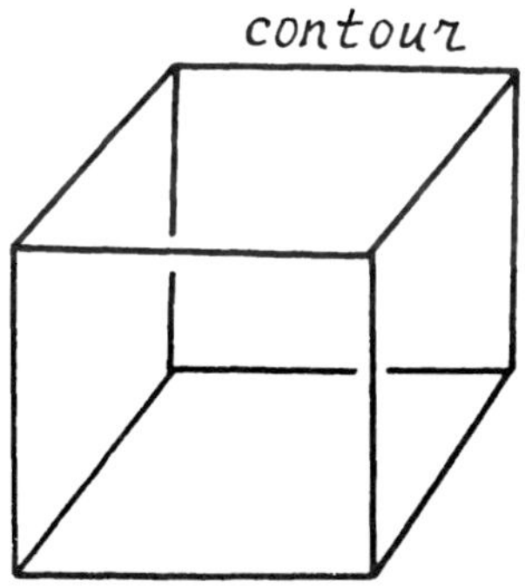

FIGURE 15F

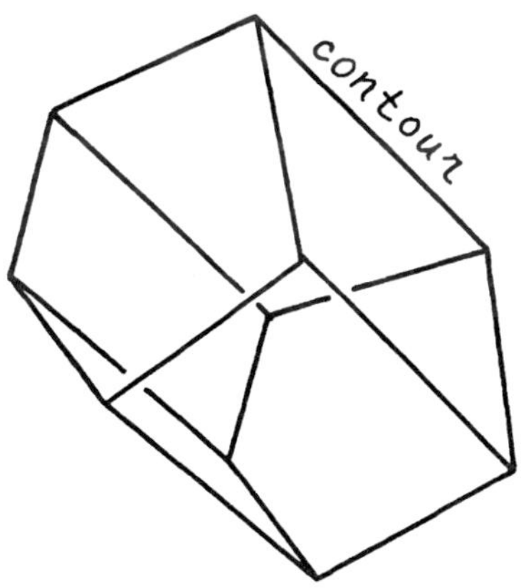

(minimal film ≠ cone)

FIGURE 15G

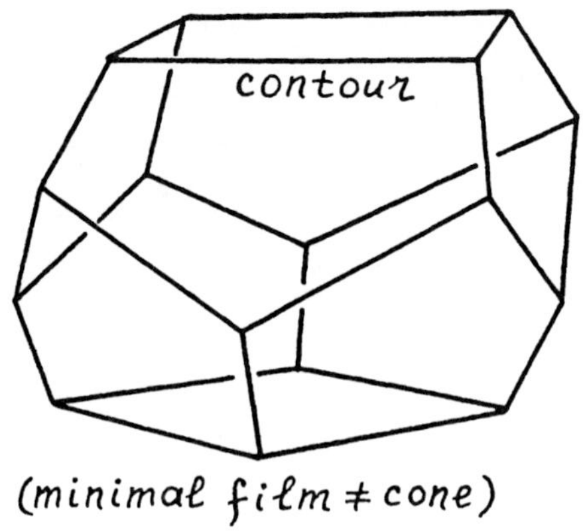

FIGURE 15H

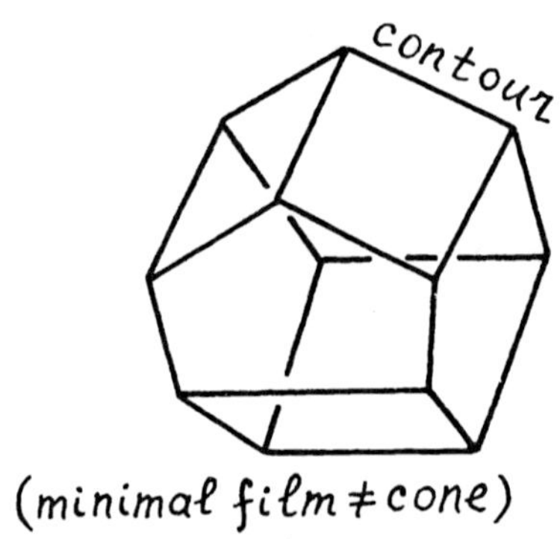

FIGURE 15I

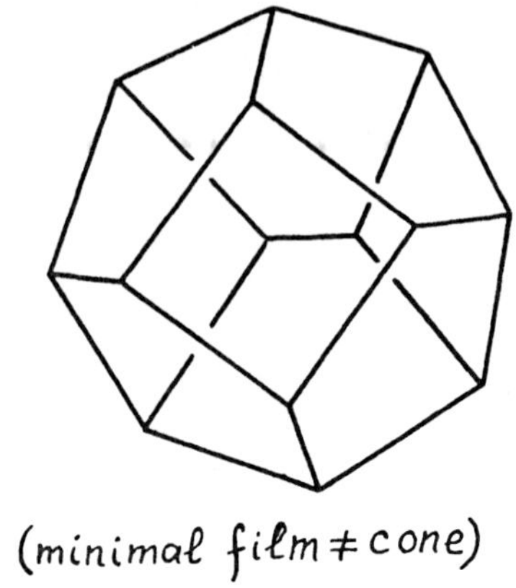

FIGURE 15J

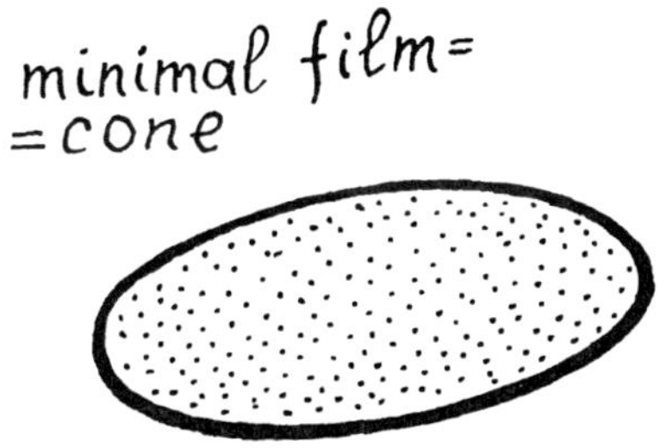

FIGURE 16A

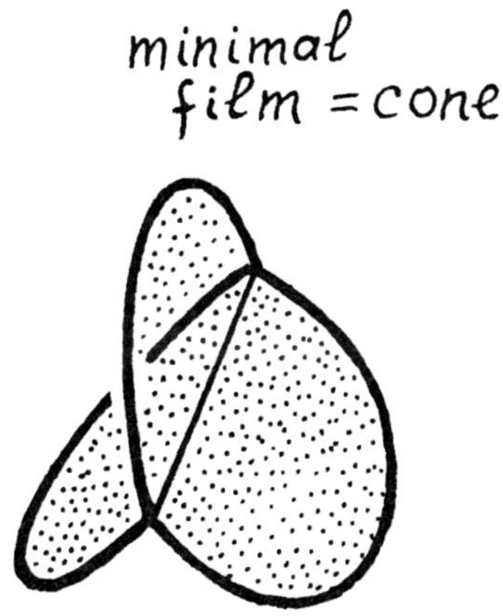

FIGURE 16B

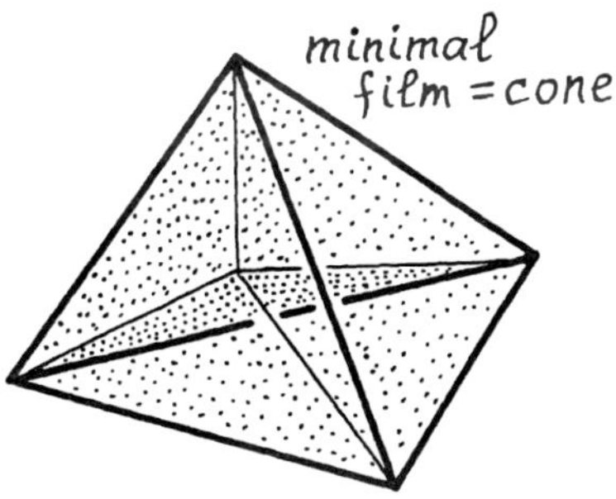

FIGURE 16C

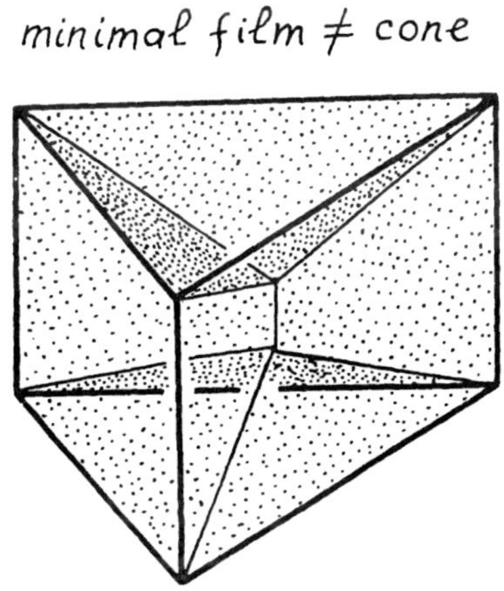

FIGURE 16D

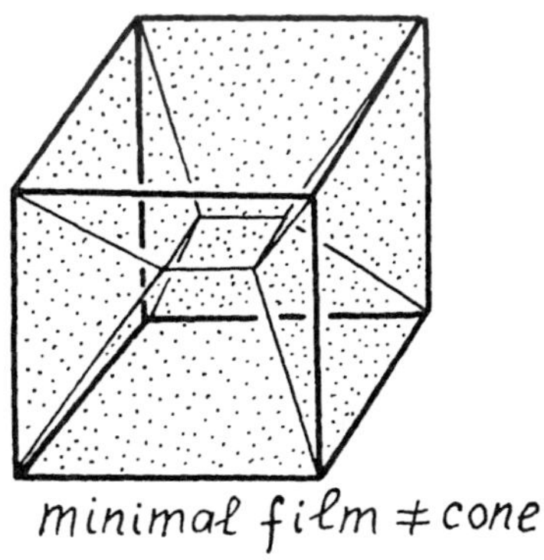

FIGURE 16E

CHAPTER 2

Classical Minimal Surfaces in $\mathbb{R}^3$

In Chapter 1 we showed how widely minimal surfaces and surfaces of constant mean curvature occur in Nature. Soap films and soap bubbles are the most convenient physical objects that enable us to obtain a visual idea of the properties of these surfaces, and experiments with a contour and soapy water often make it possible to find the right answers, suggested by Nature itself, to questions that arise in the theory of minimal surfaces and surfaces of constant mean curvature. For example, a physical experiment with soap films helps us to formulate a conjecture that gives a complete description of a bifurcation (stepwise reconstruction) of soap films spanning a Douglas contour, which arises under a continuous deformation of this contour (see [4]).

However, experimental investigation of minimal surfaces is complicated by the fact that in many cases the corresponding soap films are unstable. This means that by a small movement of the surface its area can be decreased, which leads to an instantaneous reconstruction of the corresponding soap film. The existence of unstable soap films was observed by Poisson in his investigation of the catenoid (see below). In the final section of Chapter 3 we associate with each minimal surface M an integer $\operatorname{Ind} M$, called its *index*. The index of a minimal surface determines its degree of instability, and roughly speaking it is equal to the "number of ways" of decreasing its area. If $\operatorname{Ind} M$ is not zero, then the minimal surface M is unstable. In Chapter 3 we calculate the indices of all classical minimal surfaces in $\mathbb{R}^3$ and describe some maximal domains of stability.

A theoretical investigation distinguishes among the whole class of minimal surfaces those families that have various special properties. Below we describe the main classical and modern examples of minimal surfaces known to us and say what different features they have.

§1. Catenoids

PROBLEM 1. To describe all nonplanar minimal surfaces of revolution (the so-called catenoids).

ANSWER. Each such surface is obtained by rotating a catenary (for the complete surface) or fragments of it (a catenary is the curve formed by a sagging heavy chain with uniformly distributed mass). All the complete catenoids form (up to a motion in $\mathbb{R}^3$) a one-parameter family M_a, $a > 0$: if we take the z-axis as axis of rotation, then the generator of the catenoid M_a lying in the xz-plane is given by $x = a\cosh(z/a)$ (see Figures 21 and 26).

SOLUTION. Let (r, φ, z) be cylindrical coordinates, and let M be a non-planar minimal surface of revolution about the z-axis. If $\gamma(t)$, $t \in [\alpha, \beta]$, is part of the generator of M that projects one-to-one on the z-axis, then we can parametrize the curve $\gamma(t)$ by the coordinate z. We denote the surface of revolution of γ by K. Then on K we can choose coordinates (φ, z). In these coordinates K has the representation $(r = r(z), \varphi = \varphi, z = z)$. It is easy to calculate the metric ds^2 induced on K and the area $A(K)$ of K in this metric:

$$ds^2 = dr^2 + r^2 d\varphi^2 + dz^2 = r^2 d\varphi^2 + (\dot{r}^2 + 1)\, dz^2,$$

$$A(K) = \int_K \sqrt{r^2(\dot{r}^2 + 1)}\, d\varphi\, dz = \int_\alpha^\beta 2\pi r\sqrt{\dot{r}^2 + 1}\, dz.$$

If M is minimal, then so is K. We now use the variational principle, in accordance with which a compact minimal surface is a critical point of the area functional (see §2 of Chapter 1). Consider a deformation K_s of the surface K, fixed on the boundary, in the class of surfaces of revolution (about the z-axis), where $K_0 = K$. Clearly, for sufficiently small s the surface K_s can be represented, like K, in the form $(r = r_s(z), \varphi = \varphi, z = z)$, $z \in [\alpha, \beta]$, $\varphi \in S^1$, where S^1 denotes the unit circle, the domain of variation of the coordinate φ, and $r_s(z) = r(s, z)$ is a smooth function of the variables s and z, where $r_0(z) = r(z)$. Since the deformation K_s is fixed on the boundary, all the generators $\gamma_s(z)$ of the surface K_s have the same ends $A = \gamma_s(\alpha) = \gamma(\alpha)$ and $B = \gamma_s(\beta) = \gamma(\beta)$. Moreover, the area $A(K_s)$ of K_s, which is equal to

$$A(K_s) = \int_\alpha^\beta 2\pi r_s\sqrt{\dot{r}_s^2 + 1}\, dz,$$

is a smooth function of s. For K to be minimal it is necessary that the derivative $dA(K_s)/ds|_{s=0}$ should be equal to zero for any such deformation. We use this condition to determine possible forms of generators of catenoids. It turns out that this condition is also sufficient.

We note that the area $A(K)$ of a surface of revolution K is in fact a function of all possible curves γ, the generators of the surface K: $A(K) = A(\gamma)$. The integrand $L(r, \dot{r}) = 2\pi r\sqrt{\dot{r}^2 + 1}$ depends on the coordinate r and its derivative $\dot{r}$. Formally we can regard r and $\dot{r}$ as independent variables, and $L(r, \dot{r})$ as a function of two arguments $L = L(x, y)$, where $x = r$, $y = \dot{r}$. More generally, we consider an arbitrary function $L(x, y, t)$ of three variables: $x = (x^1, \dots, x^n)$, $y = (y^1, \dots, y^n)$, and t. In the case when $L = 2\pi r\sqrt{\dot{r}^2 + 1}$ we have $n = 1$, $x = r$, $y = \dot{r}$, and the function L does not depend on t, $t = z$.

For each curve $\gamma(t)$, $t \in [\alpha, \beta]$, lying in the domain of variation of the coordinate x, we define the number $F(\gamma)$ as follows:

$$F(\gamma) = \int_\alpha^\beta L(x(t), \dot{x}(t), t)\, dt,$$

where $\dot{x}(t) = (\dot{x}^1(t), \dots, \dot{x}^n(t))$ is the velocity vector of the curve γ at the point $\gamma(t)$. We have obtained a function $F(\gamma)$ defined on all possible piecewise-smooth curves $\gamma(t)$, $t \in [\alpha, \beta]$, lying in the domain of definition of the coordinate x (we assume that the domain of variation of y is the whole of $\mathbb{R}^n$, and that of the coordinate t is the interval $[\alpha, \beta]$). The function $F(\gamma)$ is usually called a *functional*, and the function $L(x, y, t)$ is called the *Lagrangian* corresponding to the functional F.

Let $\gamma_s(t)$ be a smooth deformation of the curve γ, fixed at the ends, that is, $\gamma_s(\alpha) = \gamma(\alpha) = A$ and $\gamma_s(\beta) = \gamma(\beta) = B$, $\gamma_0(t) = \gamma(t)$. If $x_s(t)$ is the representation of the curves $\gamma_s(t)$ in the coordinates $x = (x^1, \dots, x^n)$, then, applying the functional F to each curve γ_s, we obtain a smooth function

$$F(s) = F(\gamma_s) = \int_\alpha^\beta L(x_s(t), \dot{x}_s(t), t)\, dt.$$

DEFINITION. A curve γ is called an *extremal* of a functional F if

$$dF(\gamma_s)/ds|_{s=0} = 0$$

for any deformation γ_s of γ that is fixed at the ends.

CONCLUSION. Any interval $\gamma(t)$, $t \in [\alpha, \beta]$, of a generator of a minimal surface of revolution that is projected one-to-one on the axis of rotation is an extremal of the functionals $F(\gamma) = \int_\alpha^\beta 2\pi r\sqrt{\dot{r}^2 + 1}\, dz$.

How do we find extremals of the functional F? The following assertion is an important advance in the solution of this problem.

ASSERTION 1 (see [1]). *Let F be the functional corresponding to the Lagrangian $L(x, y, t)$ defined on all possible smooth curves $\gamma(t)$, $t \in [\alpha, \beta]$, lying in a domain V. Then the curve γ is an extremal of F if and only if along the curve γ the expression $d(\partial L/\partial y)/dt - \partial L/\partial x$ is identically zero. Here*

$$\frac{\partial L}{\partial x} = \left(\frac{\partial L}{\partial x^1}, \dots, \frac{\partial L}{\partial x^n}\right), \qquad \frac{\partial L}{\partial y} = \left(\frac{\partial L}{\partial y^1}, \dots, \frac{\partial L}{\partial y^n}\right),$$

and the total derivative $d(\partial L/\partial y)/dt$ is defined as follows: in the expression $\partial L(x, y, t)/\partial y$ we need to first substitute $x = x(t)$ and $y = \dot{x}(t)$, and then differentiate the resulting function of t, that is, $\partial L(x(t), \dot{x}(t), t)/\partial y$ with *respect to* t. *Explicitly* (*in the case* $n = 1$) *we have*

$$\frac{d}{dt}\left(\frac{\partial L}{\partial y}\right) = \frac{\partial^2 L}{\partial x \partial y}\dot{x} + \frac{\partial^2 L}{\partial y^2}\ddot{x} + \frac{\partial^2 L}{\partial t \partial y}.$$

The system of differential equations

$$\frac{d}{dt}\left(\frac{\partial L}{\partial \dot{x}}\right) - \frac{\partial L}{\partial x} = 0$$

is called the *Euler-Lagrange system of equations.*

REMARK. We say that the Lagrangian L does not depend explicitly on the "time" t if $\partial L/\partial t \equiv 0$. Note that the Lagrangian need not depend explicitly on t, nevertheless dL/dt will not be equal to zero.

Before proceeding to a proof of Assertion 1, let us consider a simple example.

EXAMPLE. The motion of a point mass m under the action of the attracting force of the Earth.

Let the coordinates (x^1, x^2, x^3) be chosen in such a way that the x^3-axis is vertical and directed upwards, and the (x^1, x^2)-plane is horizontal. It is known that the force of gravity that acts on a small ball of mass m is constant and equal to $f = m\vec{g}$, where $\vec{g}$ is the acceleration of free fall, $\vec{g} = (0, 0, -g)$, where $g \approx 9.8\,\mathrm{ms}^{-2}$. Here we are considering an approximate model, well-known in the school physics course. If $v = \dot{x} = (\dot{x}^1, \dot{x}^2, \dot{x}^3)$ is the velocity of the ball, then the equation of its motion (Newton's second law) has the form

$$\frac{dp}{dt} = \frac{d(mv)}{dt} = m\vec{g} = f$$

where $p = mv$ is the impulse. Let T and U be the kinetic and potential energy of the ball. It is well known that $T = mv^2/2$ and $U = mgx^3$, where $v^2 = \sum_{i=1}^{3}(\dot{x}^i)^2$.

It turns out that the trajectories of the motion of the ball are extremals of a certain functional. In fact, we define the Lagrangian L by putting $L = T - U = mv^2/2 - mgx^3$. It is easy to verify that the impulse p and the force f can be expressed in terms of this Lagrangian as follows:

$$p = mv = \frac{\partial L}{\partial v} = \left(\frac{\partial L}{\partial v^1}, \frac{\partial L}{\partial v^2}, \frac{\partial L}{\partial v^3}\right) = \frac{\partial L}{\partial \dot{x}},$$
$$f = m\vec{g} = \frac{\partial L}{\partial x} = \left(\frac{\partial L}{\partial x^1}, \frac{\partial L}{\partial x^2}, \frac{\partial L}{\partial x^3}\right).$$

The equations of motion can be rewritten in terms of the Lagrangian L in the following form:

$$\frac{dp}{dt} = \frac{d}{dt}\left(\frac{\partial L}{\partial \dot{x}}\right) = \frac{\partial L}{\partial x} = f.$$

Thus, Newton's second law produces a system of Euler-Lagrange equations for the functional F with Lagrangian L equal to $mv^2/2 - mgx^3$. Therefore the trajectories γ of the motion of a small ball in a gravitational field are extremals of the functional $F(\gamma) = \int L\,dt$, where $L = mv^2/2 - mgx^3$.

PROOF OF ASSERTION 1. Let $\gamma_s(t) = x_s(t)$ be a deformation of the curve $\gamma(t)$, fixed at the ends $A = \gamma(\alpha) = \gamma_s(\alpha)$ and $B = \gamma(\beta) = \gamma_s(\beta)$, where $\gamma_0(t) = \gamma(t)$. We define the field η of the deformation γ_s as the vector field along the curve γ obtained as follows: with each point $t \in [\alpha, \beta]$ we associate the vector $\eta(t)$, the velocity vector of the motion of the point $\gamma(t)$ under the deformation $\gamma_s(t)$ at the initial instant $s = 0$: $\eta(t) = \partial\gamma_s(t)/\partial s|_{s=0}$. We have $F(\gamma_s) = \int_\alpha^\beta L(x_s, \dot{x}_s, t)\,dt$, so

$$\left.\frac{dF}{ds}\right|_{s=0} = \int_\alpha^\beta \left.\left[\frac{\partial L}{\partial x}\cdot\frac{\partial x_s}{\partial s} + \frac{\partial L}{\partial y}\cdot\frac{\partial \dot{x}_s}{\partial s}\right]\right|_{s=0} dt,$$

where $a\cdot b = \sum_{i=1}^n a^i b^i$. Moreover,

$$\begin{aligned}\left.\frac{dF}{ds}\right|_{s=0} &= \int_\alpha^\beta \left[\frac{\partial L}{\partial x}\cdot\eta + \frac{d}{dt}\left(\frac{\partial L}{\partial y}\cdot\eta\right) - \frac{d}{dt}\left(\frac{\partial L}{\partial y}\right)\cdot\eta\right] dt \\ &= \int_\alpha^\beta \left[\frac{\partial L}{\partial x} - \frac{d}{dt}\left(\frac{\partial L}{\partial y}\right)\right]\cdot\eta\,dt + \left.\frac{\partial L}{\partial y}\cdot\eta\right|_\alpha^\beta.\end{aligned}$$

Since the deformation γ_s is fixed at the ends, we have $\eta(\alpha) = \eta(\beta) = 0$, so $\partial L/\partial y\cdot\eta|_\alpha^\beta = 0$. Thus,

$$\left.\frac{dF(\gamma_s)}{ds}\right|_{s=0} = \int_\alpha^\beta \left[\frac{\partial L}{\partial x} - \frac{d}{dt}\left(\frac{\partial L}{\partial y}\right)\right]\cdot\eta\,dt.$$

By definition, the curve γ is an extremal if and only if $dF(\gamma_s)/ds|_{s=0} = 0$ for any deformation γ_s of the curve γ, fixed at its ends. If $\partial L(\partial L/\partial y)/\partial x - d(\partial L/\partial y)/dt$ is not equal to zero at some point t_0 of the interval $[\alpha, \beta]$, then from the continuity it follows that this expression is not equal to zero in some neighborhood of t_0. It is easy to construct a deformation, fixed outside this neighborhood and having a nonzero deformation field, that has the same directions as the field $\partial L(\partial L/\partial y)/\partial x - d(\partial L/\partial y)/dt$. For such a deformation $dF(\gamma_s)/ds|_{s=0} > 0$. This completes the proof of Assertion 1.

We note that the Euler-Lagrange equations are a system of ordinary differential equations of the second order (in the case $n = 1$ this system consists of a single equation). It turns out that in some cases the Euler-Lagrange equations can be much simplified. For example, when $n = 1$ from a second-order equation we can go over to a one-parameter family of first-order equations, whose solutions can as a rule be obtained much more easily. Before formulating the corresponding assertion, we again consider the example of the motion of a point mass in the force field of the Earth's attraction.

We recall that the total energy H of a moving ball is equal to the sum of the kinetic and potential energies, that is, $H = T + U = mv^2/2 + mgx^3$. The well-known law of conservation of energy asserts that along a trajectory of the motion the total energy is constant: $H = \text{const}$. This equation is of the first order and carries easily extracted additional information.

Let us calculate the total energy H in terms of the Lagrangian:

$$H = T + U = 2T - L = p \cdot v - L = \frac{\partial L}{\partial \dot{x}} \cdot \dot{x} - L.$$

ASSERTION 2 (the law of conservation of energy; see [1], [38]). *Let F be the functional corresponding to the Lagrangian L (see Assertion 1). Suppose that L does not depend explicitly on the "time" t, that is, $L = L(x, y)$, $\partial L/\partial t \equiv 0$. Then on an extremal γ of the functional F the total energy H, calculated from the formula $H = \partial L/\partial y \cdot y - \partial L/\partial x$, is constant, that is, $H(\gamma(t))$ does not depend on t, $H(\gamma(t)) = \text{const}$.*

PROOF.

$$\frac{dH}{dt} = \frac{d}{dt}\left(\frac{\partial L}{\partial \dot{x}}\right)\dot{x} + \frac{\partial L}{\partial \dot{x}}\ddot{x} - \frac{\partial L}{\partial x}\dot{x} - \frac{\partial L}{\partial \dot{x}}\ddot{x} = \left[\frac{d}{dt}\left(\frac{\partial L}{\partial \dot{x}}\right) - \frac{\partial L}{\partial x}\right] \cdot \dot{x} = 0.$$

REMARK. In the case when $n = 1$, that is, $x = x^1$, $y = y^1$, the law of conservation of energy $H = \text{const}$ produces a one-parameter family (parametrized by a constant) of ordinary differential equations of the first order. We now use this remark to determine all generators of minimal surfaces of revolution.

Thus, suppose that the curve $\gamma(z)$, $z \in [\alpha, \beta]$, is a generator of a minimal surface of revolution K that projects one-to-one on the z-axis (see above). The Lagrangian $L = 2\pi r\sqrt{\dot{r}^2 + 1}$ corresponding to the area functional $A(K)$ of the surface K does not depend explicitly on the "time" z. Therefore along an extremal the total energy H is conserved. We have

$$p = \frac{\partial L}{\partial \dot{r}} = \frac{2\pi r\dot{r}}{\sqrt{\dot{r}^2 + 1}},$$

$$H = p \cdot v - L = \frac{2\pi r\dot{r}^2}{\sqrt{\dot{r}^2 + 1}} - 2\pi r\sqrt{\dot{r}^2 + 1} = -\frac{2\pi r}{\sqrt{\dot{r}^2 + 1}} = a_1.$$

Consequently,

$$\dot{r} = \pm\sqrt{\frac{r^2}{a^2} - 1}, \quad a > 0; \qquad z = \pm a \operatorname{arccosh}\frac{r}{a} + c.$$

Thus, up to a shift along the z-axis and the direction of parametrization (the coordinate z or $-z$) we have obtained the equation $r/a = \cosh(z/a)$, $z \in [\alpha, \beta]$, that specifies the surface of revolution K: the generator $\gamma(z)$ of any minimal surface of revolution about the z-axis satisfies this equation for some a (after a suitable motion). We show that for any $a > 0$ the surface K is minimal (we have not verified the condition of extremality of the area functional for all deformations). For this it is obviously sufficient to verify that the mean curvature of K is zero at all points of the generator γ.

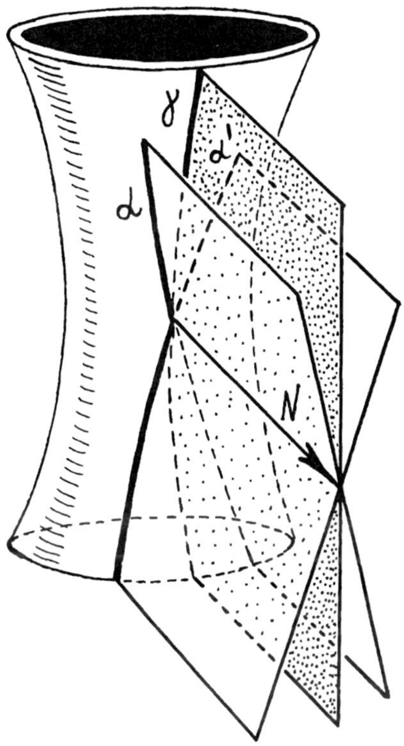

FIGURE 17

PROBLEM 2. Prove that the generator γ of a surface of revolution is a principal section for any point of it.

SOLUTION. If this is not so, then the normal section perpendicular to γ is not a principal section. Therefore for a principal section α there is a section α', not coinciding with it, that is symmetrical about the plane specifying γ. But then the sections α and α' have the same curvature, which contradicts Euler's formula (Figure 17).

LEMMA (Meusnier's theorem; see [1], and also Chapter 3). *Let M be a smooth surface in $\mathbb{R}^3$, and P a point of M. Let Π be a plane passing through P and not touching M at this point. Let v denote the unit tangent vector of the section $\gamma = \Pi \cap M$ at P, and let φ denote the angle between the plane Π and the normal plane passing through P parallel to v. Then the curvature κ of the section γ and the curvature $\kappa(v)$ of the normal section in the direction v are connected by the relation $\kappa \cos\varphi = \kappa(v)$ (see Figure 18). Here the curvature κ is calculated with respect to the normal n in the plane Π to the section γ, which forms an acute angle with the normal N to the surface M.*

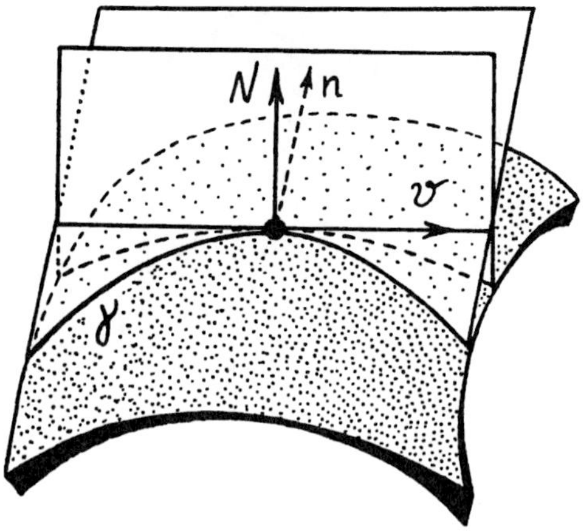

FIGURE 18

It is now easy to calculate the principal curvatures of a surface of revolution. Suppose that P lies on the generator $\gamma(z) = \{x = r(z), y = 0\}$. Then one principal curvature at P is the curvature of γ. Therefore $\kappa_1 = \ddot{r}/(1+\dot{r}^2)^{3/2}$.

The curvature κ_2 of the orthogonal normal section is calculated by means of Meusnier's theorem. The section by the plane Π orthogonal to the z-axis is a circle with curvature $\kappa = -1/r$ (the minus sign appears because of the choice of the normal N to the surface of revolution M). The cosine of the angle between the plane Π and the corresponding normal plane is equal to

$$\cos\varphi = (1, \dot{r})\frac{1}{\sqrt{1+\dot{r}^2}} \cdot (1, 0) = \frac{1}{\sqrt{1+\dot{r}^2}}.$$

Therefore $k_2 = -1/(r\sqrt{1+\dot{r}^2})$, and the mean curvature H is equal to

$$H = \frac{\ddot{r}}{(1+\dot{r}^2)^{3/2}} - \frac{1}{r\sqrt{1+\dot{r}^2}} = \frac{\ddot{r}r - \dot{r}^2 - 1}{r(\sqrt{1+\dot{r}^2})^3}.$$

The rest of the verification is obtained automatically. Since each solution can be extended to the whole z-axis, the case when the generator of a minimal surface of revolution does not project one-to-one on the z-axis anywhere leads to a flat surface (that is, this generator is not a graph over the z-axis anywhere). This completes the solution of Problem 1.

Thus, all nonplanar noncongruent complete minimal surfaces of revolution form a one-parameter family of catenoids $r = a\cosh(z/a)$, $a > 0$.

A compact catenoid can be realized as a soap film covering two coaxial circles of the same radius lying in parallel planes. How many soap films can be spanned on this contour?

In 1983 Schoen [5] showed that on such a contour we can span only minimal surfaces of revolution (Schoen's results refer to a more general situation).

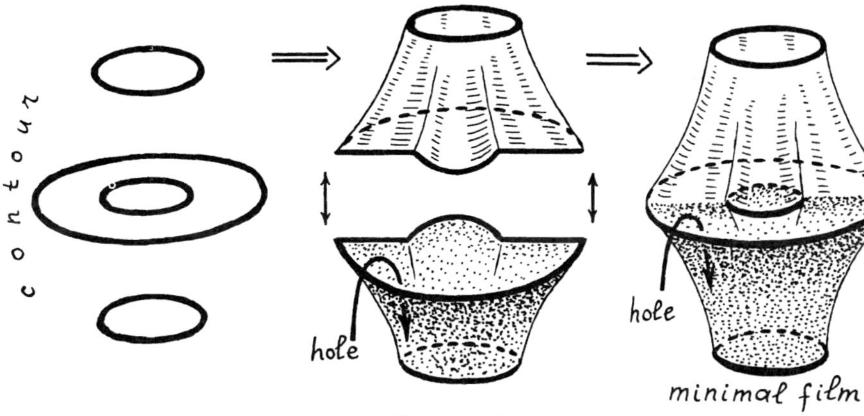

FIGURE 19A

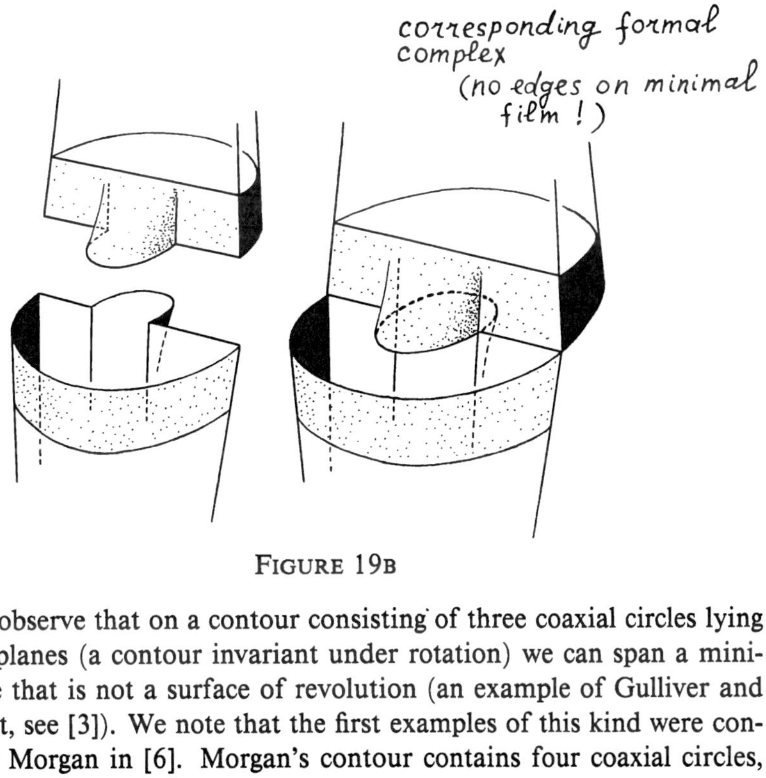

FIGURE 19B

We should observe that on a contour consisting of three coaxial circles lying in parallel planes (a contour invariant under rotation) we can span a minimal surface that is not a surface of revolution (an example of Gulliver and Hildebrandt, see [3]). We note that the first examples of this kind were constructed by Morgan in [6]. Morgan's contour contains four coaxial circles, two of which lie in the plane $z = 0$, and the other two in the planes $z = \pm 1$. Figures 19 and 20 show the contours and the corresponding minimal surfaces from the examples of Morgan and Gulliver-Hildebrandt. The main idea of both constructions is the following. To a contour that is invariant under rotation we add a number of segments such that the generating system can be split into two symmetric parts Γ_1 and Γ_2 not invariant under rotation. Covering Γ_1 and Γ_2 by symmetric soap films M_1 and M_2 and combining Γ_1 and Γ_2 into the original system, we obtain a minimal surface spanning

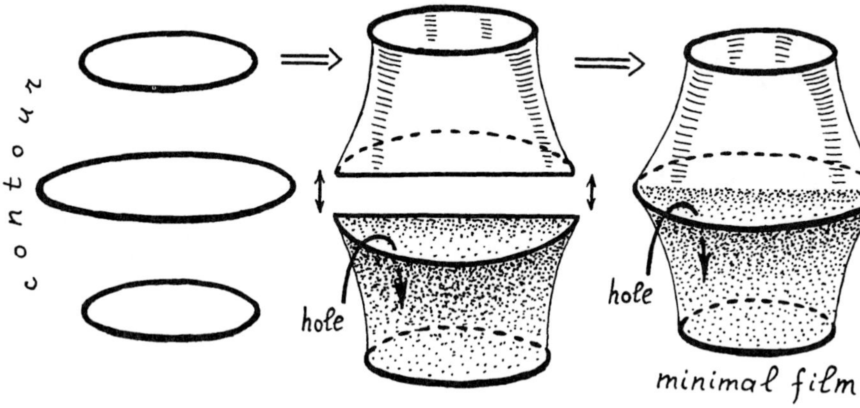

FIGURE 20

the original contour and, in accordance with the symmetry principle (see §2), is smooth along the added segments, which can be removed without losing anything. The resulting film $M_1 \cup M_2$ will not be invariant under rotation.

Thus, to describe all minimal surfaces spanning a contour that consists of two coaxial circles of the same radius situated in parallel planes it is sufficient to describe all catenoids spanning this contour. The complete solution of this problem had already been obtained by Poisson (see [2]). We now give the corresponding results.

Let h be the distance between the circles, and ρ their radii. It turns out that for small h there are exactly two catenoids $r = a_i \cosh(z/a_i)$, $i = 1, 2$; one of them is close to a cylinder, and the other to a cone (see Figure 21(a)). As $h \to 0$ the radius a_1 of the mouth of the first catenoid tends to the radius of the bounding circle, $a_1 \to \rho$, and the radius a_2 of the mouth of the second catenoid tends to zero, $a_2 \to 0$. If h increases, then the outer catenoid sags (a_1 decreases), and conversely the inner catenoid becomes less steep (a_2 increases). Thus, as h increases the catenoids tend to each other. Finally, for some value $h = h_{\text{cr}}(\rho)$ the two catenoids "stick together". On such a contour we can span only one catenoid. If we continue to move the circles apart, then in practice the catenoid corresponding to $h_{\text{cr}}(\rho)$ suddenly becomes the film spanning each flat disk of the circles: when $h > h_{\text{cr}}$ there is no catenoid spanning this contour (see Figure 21(b)).

For an analytical justification of the behavior of the catenoids as we vary the distance between the circles, assume that in a cylindrical coordinate system (r, φ, z) the circles of radius ρ forming our contour lie in the planes $z = \pm\, h/2$ and the line of centers is the z-axis. Then the catenoids spanning this contour have the form $r = a\cosh(z/a)$ with $\rho = a\cosh(h/2a)$. The number of catenoids is equal to the number of solutions of the last equation for the variable a, $a > 0$.

We construct the graph $y = a\cosh(h/2a)$ in coordinates (a, y) for some fixed h. The required solutions are given by the intersection of this graph with the line $y = \rho$. As $a \to 0$ or $a \to +\infty$ we obviously have $y \to +\infty$.

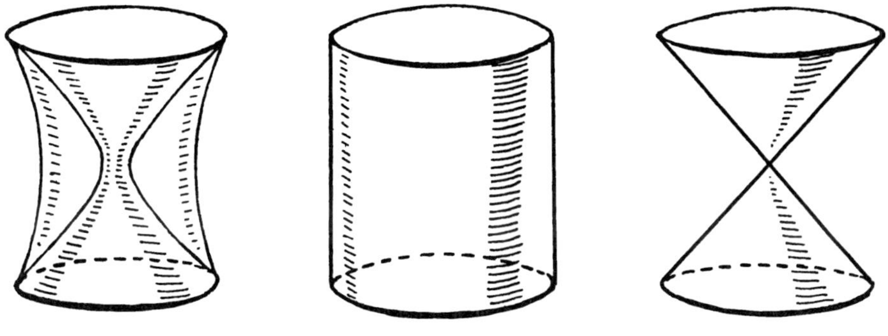

FIGURE 21A

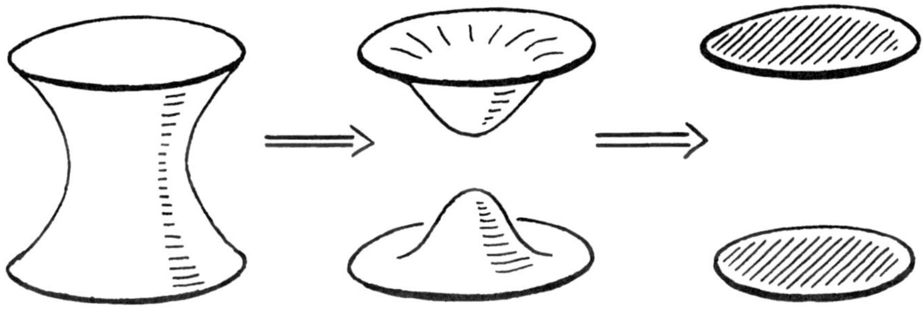

FIGURE 21B

We show that when $a > 0$ the function $y(a) = a\cosh(h/2a)$ has a unique extremum (a minimum). In fact, the equation

$$\frac{dy}{da} = \cosh\frac{h}{2a} - \frac{h}{2a}\sinh\frac{h}{2a} = 0$$

has a unique positive solution $a = a_0$, since this is true for the equivalent equation $\operatorname{cotanh}(h/2a) = h/2a$, $a > 0$. Let t_0 denote the unique positive root of the equation $\operatorname{cotanh} t = t$. Then $a_0 = a_{cr} = h/2t_0$. Thus, the graph for $y(a) = a\cosh(h/2a)$ is as shown in Figure 22.

If we now begin to increase h, the graph moves upwards. As $h \to +\infty$ the critical point $a_0 = h/2t_0 \to +\infty$, and so the critical value $y_0 = a_0\cosh(h/2a_0) = a_0\cosh t_0 \to +\infty$. Conversely, as $h \to 0$ we have $a_0 \to 0$ and $y_0 \to 0$.

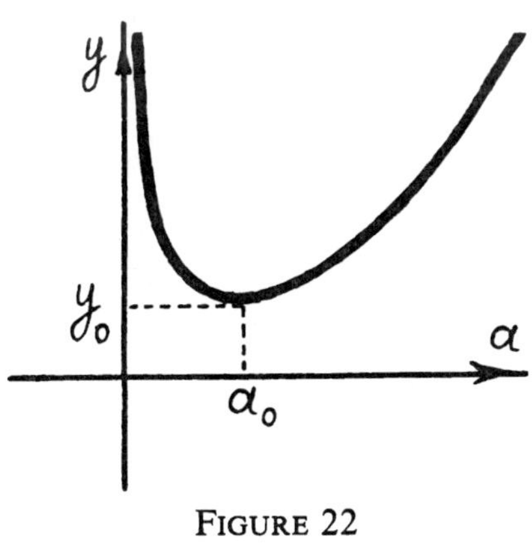

FIGURE 22

CONCLUSION. For any $\rho > 0$ there is a critical value $h_{cr}(\rho)$ such that when $h < h_{cr}(\rho)$ the line $y = \rho$ intersects the graph of the function $y(a) = a\cosh(h/2a)$ in exactly two points which tend to each other as $h \to h_{cr}(\rho)$, and when $h = h_{cr}(\rho)$ the line $y = \rho$ touches the graph for $y(a) = a\cosh(h/2a)$. When $h > h_{cr}(\rho)$ there are no intersections. This justifies the geometrical picture described above of the behavior of the catenoids as h varies.

We note that in real experiments we always obtain only one of the two catenoids, namely the one that is close to a cylinder for small h. The fact is that the second catenoid is unstable (see Chapter 3).

§2. The helicoid

A helicoid is a surface swept out by a straight line (the generator of the helicoid) under a uniform screw motion (see Figure 23).

In order to give an exact meaning to this definition, let us write down the parametric form of a helicoid. We introduce Cartesian coordinates x, y, z in such a way that the z-axis coincides with the axis of rotation and the x-axis with the generator of the helicoid. If h denotes the distance between the closest noncoincident parallel generators (h is called the *pitch* of the helicoid), v is the angle between the generator and the xz-plane, and u is the standard coordinate on the generator (points with $u = 0$ lie on the z-axis), then the helicoid can be defined parametrically as

$$x = u\cos v\,, \qquad y = u\sin v\,, \qquad z = av\,,$$

where $|a| = h/2\pi$, $u \in \mathbb{R}$, and $v \in \mathbb{R}$.

If $a > 0$, then an anticlockwise rotation of the generator raises it (z increases), and if $a < 0$ the rotation lowers it (z decreases).

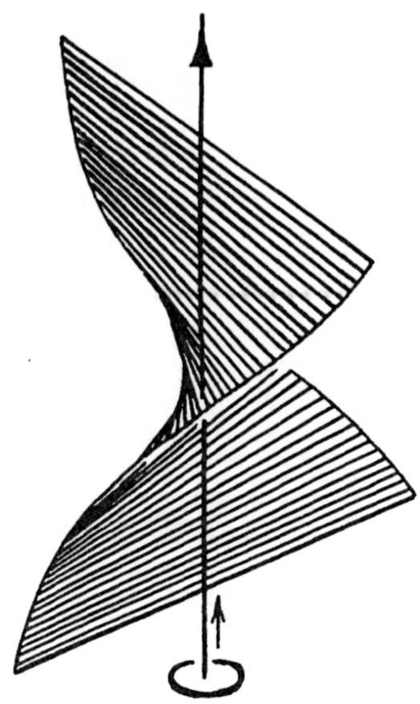

FIGURE 23

A simpler representation of the helicoid can be obtained in cylindrical coordinates (r, φ, z), but in this case only "half" the helicoid is parametrized: if for u and v we take $u = z$, $v = \varphi$, then "half" the helicoid can be given by $(r = r, \varphi = \varphi, z = a\varphi)$, or simply $z = a\varphi$.

The helicoid belongs to the family of ruled surfaces, that is, surfaces swept out by a straight line (the generator of the ruled surface) as it moves along some curve, called the *directrix* of the ruled surface. All cylinders (the generators are all parallel) and cones (the generators all pass through one point) are ruled surfaces. An interesting example is a ruled surface obtained by rotating the generator about an axis of rotation skew to it. If the generator and the axis of rotation are not perpendicular, this surface is called a hyperboloid of one sheet and is an example of a second-order surface (the equation specifying the hyperboloid is a polynomial of the second degree in the Cartesian coordinates x, y, z). We note that a plane, which is also a ruled surface, is of both conical and cylindrical type. Moreover, a plane can be regarded as a degenerate helicoid, having zero pitch $h = 0$.

It turns out that a helicoid is a minimal surface. Moreover, among all ruled surfaces only the helicoid and fragments of it are minimal.

This assertion was proved by E. Catalan in 1842. Here we give a different proof, which is more geometrical from our point of view.

We first prove that a helicoid is a minimal surface. For this it is sufficient to verify that the mean curvature along an arbitrary generator, say the x-axis, is zero, since for any pair of generators there is a motion that takes the helicoid into itself and the first generator into the second.

Suppose that the point P of the generator X (the x-axis) has coordinates $(u, 0, 0)$. Through P we draw the plane Π perpendicular to X. Clearly, the intersection of Π with the helicoid is a normal section, which we denote by γ. The plane Π is given by the equation $x - u = 0$. The normal section γ obviously has the form $\gamma(\varphi) = (u, u\tan\varphi, a\varphi)$, where $\gamma(0) = P$, so the curve γ in the plane Π is the graph of the tangent function, which has a point of inflexion at $\varphi = 0$. This means that the curvature of γ at P is zero. Thus, for two mutually perpendicular normal sections—the generator X and the curve γ—the curvature is zero (at P). From this it follows immediately that the mean curvature H vanishes at P (see the problem in §1 of Chapter 1). The case $u = 0$ is obtained similarly. Since P is arbitrary, we have proved that the helicoid is a minimal surface.

To prove that the helicoid is the only complete ruled minimal surface we need two properties of minimal surfaces, namely the reflection principle of Schwarz and Riemann and the uniqueness theorem. Both these properties follow from the fact that in the special coordinates (u, v) the functions $x(u, v)$, $y(u, v)$, $z(u, v)$ that specify a minimal surface parametrically are harmonic (see §2 of Chapter 3). This remark enables us to extend many properties of harmonic and analytic functions to minimal surfaces.

THE SCHWARZ-RIEMANN REFLECTION PRINCIPLE 1. Suppose that the boundary of a minimal surface M lying in $\mathbb{R}^3$ contains an interval I of a straight line l. Let M^* denote the minimal surface symmetrical to M about the line l. Then $M^* \cup M$ is a smooth minimal surface: the surfaces M and M^* join smoothly along the interval I (compare with the symmetry principle for analytic and harmonic functions).

THE UNIQUENESS THEOREM. *If two smooth minimal surfaces M_1 and M_2 contain an open subset in their intersection, then the union $M_1 \cup M_2$ of these surfaces is a smooth minimal surface (compare with the uniqueness theorem for analytic and harmonic functions).*

We now give an outline of the proof of Catalan's theorem.

It turns out that ruled minimal surfaces are completely characterized by the following property.

LEMMA. *Let M be a ruled minimal surface with directrix $\gamma(t)$, and let l_0 and l_1 be the generators of M passing through the points $\gamma(t_0)$ and $\gamma(t_1)$ respectively. Then there is a generator l passing through the point $\gamma(t)$ for some t, $t_0 < t < t_1$, such that l_0 is symmetrical to l_1 about l (we call such a line l the midline between l_0 and l_1).*

PROOF. Consider an arbitrary generator l passing through some point $\gamma(t')$, $t_0 < t' < t_1$. Let $M[t_0, t']$ denote the part of the surface M consisting of all generators drawn through the point $\gamma(t)$ for all $t \in [t_0, t']$. Let us reflect $M[t_0, t']$ about l. We denote the image by $M^*[t_0, t']$. In accordance with the reflection principle and the uniqueness theorem the union $M[t_0, t'] \cup M^*[t_0, t']$ is either equal to $M[t_0, t^*]$ for some t^*, $t_0 < t^* < t_1$, or it contains $M[t_0, t_1]$. As $t' \to t_0$, t^* also tends to t_0. As $t' \to t_1$, t^* at some instant becomes greater than t_1. Because of the continuity it follows that there is a t' such that $t^* = t_1$. This proves the lemma.

Next, we define a frame of an arbitrary ruled surface as a family of generators that is everywhere dense (on the surface). Obviously the frame completely determines the form of the ruled surface: if two ruled surfaces M_1 and M_2 have congruent frames, then the surfaces themselves are congruent.

Using the characteristic property of a ruled minimal surface described in the lemma, on each such surface we construct a frame, which, as it turns out, is congruent either to a planar frame or a frame lying on the helicoid. This observation completes the proof of Catalan's theorem.

Thus, let M be an arbitrary ruled minimal surface with directrix $\gamma(t)$, and let l_0 and l_1 be an arbitrary pair of noncoincident generators of M passing through the points $\gamma(t_0)$ and $\gamma(t_1)$ respectively. We draw the midline $l = l_{1/2}$ between l_0 and l_1. From the lemma it follows that l is a generator of M. Next, let $l_{1/4}$ be the midline between l_0 and $l_{1/2}$, and let $l_{3/4}$ be the midline between $l_{1/2}$ and l_1. By induction, if l_0, $l_{1/2^n}$, $\ldots$, l_1 is the family of generators constructed at the nth stage, then at the $(n+1)$st stage we add to this family all the midlines drawn between neighboring generators $l_{j/2^n}$ and $l_{(j+1)/2^n}$ for some j, $j = 0, 1, \dots, 2^n - 1$. In the limit we obviously obtain a frame L of the surface $M[t_0, t_1]$. To obtain a frame for the whole surface M, we reflect L about the lines l_0 and l_1 and consider the union of the images with L itself. The resulting collection is a frame of the ruled minimal surface $M_{-1,2} = M^*_{-1} \cup M[t_0, t_1] \cup M^*_2$, where M^*_{-1}, and M^*_2 are the images of the surface $M[t_0, t_1]$ under reflection about the generators l_0 and l_1 respectively. Continuing the reflection process in the obvious way, in the limit we obtain a frame L_∞, the frame of the complete ruled minimal surface, which, in accordance with the uniqueness theorem, contains our minimal surface M.

It remains to show that the frame L_∞ is either planar or the frame of a helicoid. In fact, if l_0 and l_1 are parallel or intersect, then L_∞ is obviously the frame of a plane (all the midlines lie in one plane). Suppose that l_0 and l_1 are skew. Consider a plane Π parallel to the two generators and project l_0 and l_1 orthogonally onto Π. Let l'_0 and l'_1 be their projections, and let $P \in \Pi$ be the point of intersection of l'_0 and l'_1. If l is the midline between l_0 and l_1, then the orthogonal projection l' of l on Π is the bisector of the angle between l'_0 and l'_1. Moreover, l is at the same distance from l_0 and l_1. From what we have said it follows that if we draw through P the line Z perpendicular to Π, then firstly the lines l_0, l, and l_1 are orthogonal to Z, secondly they intersect Z in points z_0, z, and z_1 such that $|z_0 z| = |z z_1|$, and thirdly the angle between l_0 and l is equal to the angle between l and l_1. Therefore l lies on some helicoid (with Z as directrix) passing through l_0 and l_1. Similar results hold for any triple of lines of L_∞, one of which is the midline between the other two. It is now clear that L_∞ is the frame of a helicoid. This completes the proof of Catalan's theorem.

We have thus shown that all nonplanar noncongruent complete minimal ruled surfaces form a one-parameter family of helicoids (for the parameter of the family we can choose the pitch h). We recall that all nonplanar noncongruent complete minimal surfaces of revolution also form a one-parameter family, the family of catenoids. For the parameter of this family we can take the radius of the mouth of the catenoid. It turns out that there is a very close connection between catenoids and helicoids. Firstly, the catenoid and helicoid are locally isometric for suitable values of the parameters. This means the following. We recall that in cylindrical coordinates (r, φ, z) the catenoid is given by $r = a\cosh(z/a)$, $a > 0$, where $z \in \mathbb{R}$, and the angle φ varies from 0 to 2π. If we allow the angle φ to vary from $-\infty$ to $+\infty$, we obtain a covering, winding of a catenoid with infinitely many sheets. It turns out that for suitable values of the parameters h and a there is a diffeomorphism between the helicoid and this covering of a catenoid that preserves the metric.

Moreover, by means of a bending we can wind the helicoid onto the corresponding catenoid, where the bending can be carried out in the class of minimal surfaces. This means that there is a one-parameter family of minimal surfaces M_t, $0 \le t \le 1$, smoothly depending on the parameter t, such that M_0 is a helicoid, M_1 is a covering of a catenoid with infinitely many sheets, and all the surfaces M_t are isometric to one another. Below we construct this bending explicitly. Isometric minimal surfaces M_0 and M_1 that can be "joined" by a smooth family M_t of isometric minimal surfaces are said to be *conjugate*, and the intermediate minimal surfaces M_t, $0 < t < 1$, are said to be *associated*. In §3 of Chapter 3 we give another definition (more convenient for our work) of an associated family and conjugate surfaces in terms of the Weierstrass representation. For now we just observe that the catenoid and helicoid (the case $a = 1$, $h = 2\pi$) can be given parametrically as follows:

$$\text{catenoid} \quad r_1(u, v) = \begin{pmatrix} x \\ y \\ z \end{pmatrix} = \operatorname{Re} \begin{pmatrix} \cosh w \\ -i \sinh w \\ w \end{pmatrix} = \begin{cases} \cosh u \cos v, \\ \cosh u \sin v, \\ u, \end{cases}$$

$$\text{helicoid} \quad r_2(u, v) = \begin{pmatrix} x \\ y \\ z \end{pmatrix} = \operatorname{Im} \begin{pmatrix} \cosh w \\ -i \sinh w \\ w \end{pmatrix} = \begin{cases} \sinh u \sin v, \\ -\sinh u \cos v, \\ v, \end{cases}$$

where (u, v) are coordinates on the surface, and $w = u + iv$ is the corresponding complex coordinate, $w \in C$. We note that the parameter v is the angle φ in cylindrical coordinates, and for the helicoid the coordinate u is replaced by $\sinh u$. The joining family of minimal surfaces is given by the radius vector $r(u, v, \alpha)$ as follows:

$$r(u, v, \alpha) = \cos a \cdot r_1(u, v) + \sin a \cdot r_2(u, v),$$

where the parameter $\alpha \in [0, \pi/2]$. When $\alpha = 0$ we have a covering of a catenoid, and when $\alpha = \pi/2$ we have a helicoid. The fact that all these surfaces are minimal follows immediately from a result in §3 of Chapter 3. The fact that they are all isometric is obtained by direct calculation (verify this).

Let us fix some $\alpha \in (0, \pi/2)$ and see what the surface $r(u, v, \alpha)$ looks like. We need to find the "sum" of a catenoid with mouth radius $\cos\alpha$ and a helicoid with pitch $2\pi \sin\alpha$. Let us put $v = 0$ and see how the corresponding generators of the catenoid and helicoid are situated.

The generator of the catenoid is a catenary $\gamma(u)$ in the xz-plane; as u increases the z-coordinate of the point $\gamma(u)$ increases. The generator of the helicoid is the y-axis; as u increases the y-coordinate decreases. We need to find the "sum" ω of the radius vectors of these generators for the same values of u. In order to represent ω visually, we draw through the catenary γ the family of lines parallel to the y-axis. We obtain a cylindrical surface. Clearly, to the value $u = 0$ there corresponds the vertex of γ. As u increases the corresponding point of ω moves with γ along the cylindrical

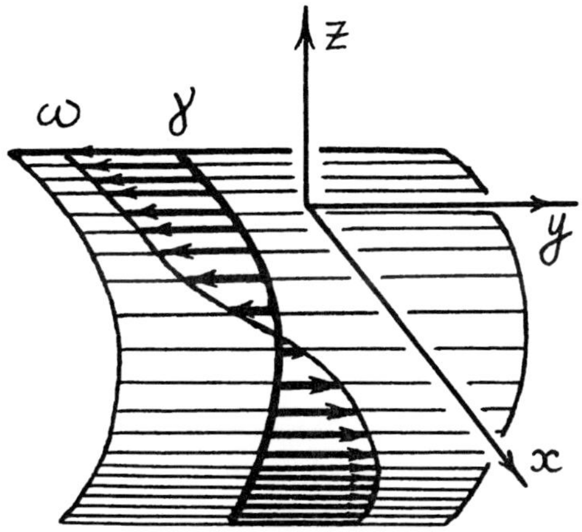

FIGURE 24

surface to the left, and as u decreases it moves to the right. The larger $|u|$ is, the larger the shift (see Figure 24).

What happens to ω if we raise the generator of the helicoid by a height h? Obviously ω is also raised by a height h. One final remark. As v increases, the generators of the catenoid and helicoid rotate anticlockwise about the z-axis with the same speed. The generator of the helicoid moves forward in the direction of the z-axis. Thus, the surface $r(u, v, \alpha)$ is rather like a helicoid: it is obtained as the surface swept out by the curve ω under a simultaneous uniform rotation about the z-axis and a uniform forward motion along this axis.

Let us see what happens to the "generator" ω as α varies from 0 to $\pi/2$. When $\alpha = 0$ the curve ω is a catenary γ of the catenoid $r_1(u, v)$. As α increases the vertex of γ tends to the origin, and ω deviates more and more from γ along the corresponding cylindrical surface. The profiles of different cylindrical surfaces for different α are shown in Figure 25.

In the limit, when $\alpha = \pi/2$, the curve ω becomes horizontal, and its vertex coincides with the origin—ω becomes a generator of the helicoid. The distance between neighboring coils of the surface $r(u, v, \alpha)$ is equal to

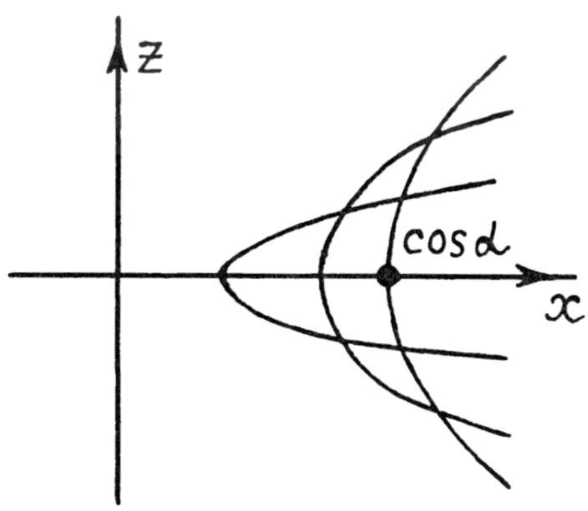

FIGURE 25

$2\pi \sin\alpha$, so it tends to 2π as $\alpha \to \pi/2$. The deformation of the helicoid into a catenoid is shown visually in Figure 26.

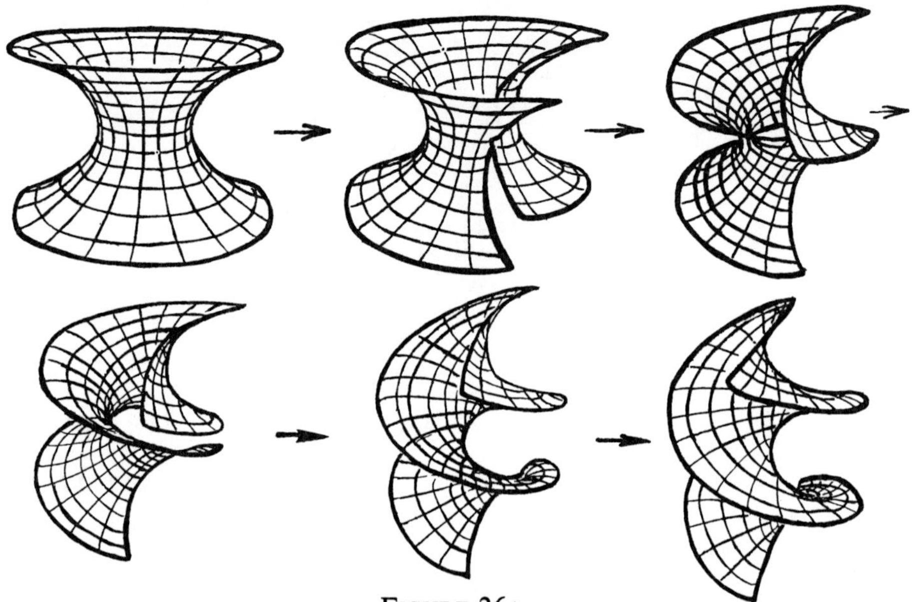

FIGURE 26A

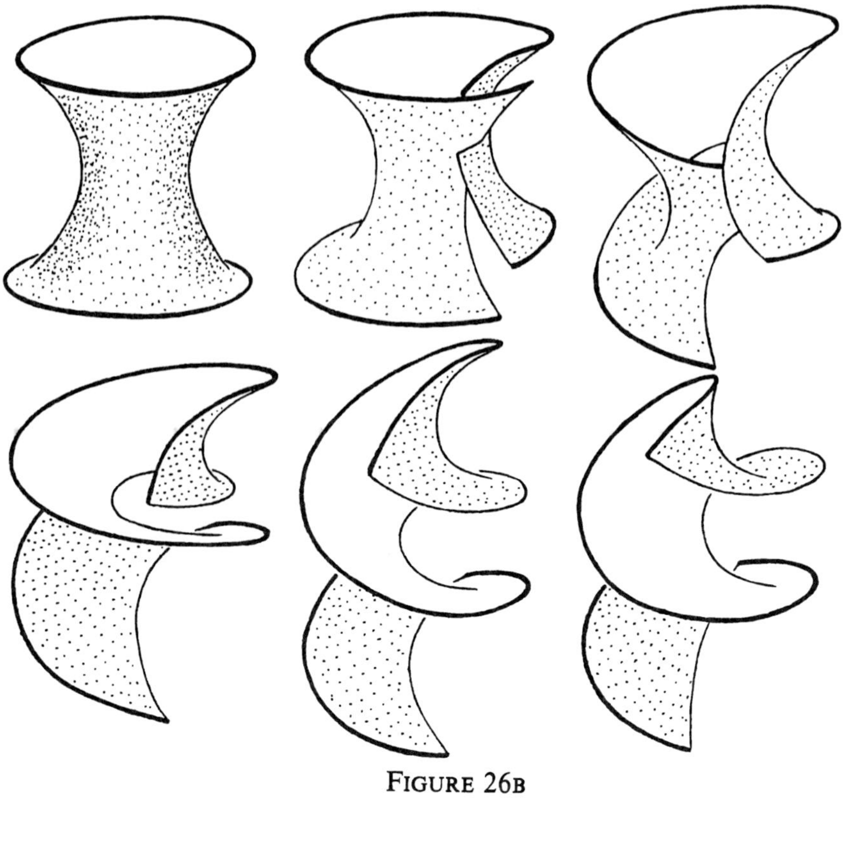

FIGURE 26B

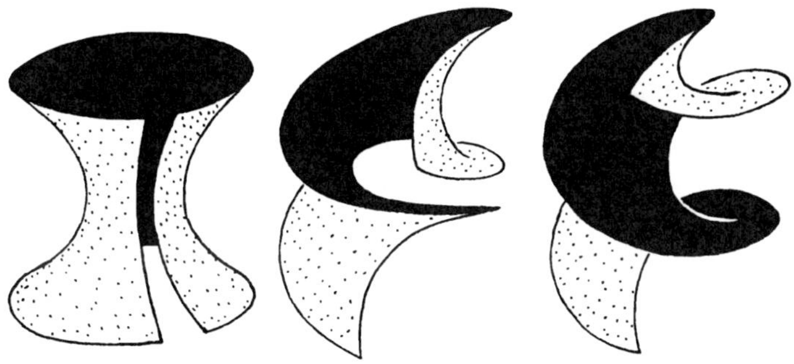

FIGURE 27A

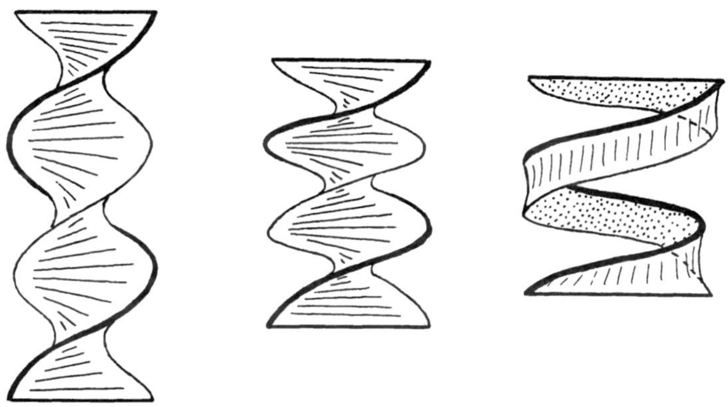

FIGURE 27B

We note that a similar deformation can easily be seen by experiment with soap films. For this we make a wire contour formed from two cutting circles and two connecting wires going along closing meridians of the catenoid (see Figure 27(a)). Deforming this contour, as shown in the figure, we obtain what is required.

Interesting experiments can be set up by investigating the helicoid for stability. We make a contour consisting of two spirals and two closing intervals. For large pitch (that is, for sufficiently large period of the spirals) a soap film stretched on this contour is a physical realization of a helicoid. If we decrease the pitch, contracting the spirals, there is an instant when the film ceases to be a helicoid and becomes a film of type (2) (see Figure 27(b)). In Chapter 3 we calculate the index of the helicoid and prove that it is an unstable minimal surface.

§3. The minimal surface equation. Bernstein's problem. The Scherk surface

1. The minimal surface equation in $\mathbb{R}^3$. Suppose that a minimal surface M in $\mathbb{R}^3$ is given in the form of the graph of a function $z = f(x, y)$ over a domain $\Omega \subset \mathbb{R}^2_{(x,y)}$. How can we describe all such functions f? It turns out that all functions whose graphs are minimal surfaces are described by the following second-order differential equation:

$$(1+f_y^2)\cdot f_{xx} - 2f_{xy}f_xf_y + (1+f_x^2)\cdot f_{yy} = 0.$$

This is called the *minimal surface equation.*

For the proof we use the fact that minimal surfaces are critical points of the area functional (see §2 of Chapter 1). Let Ω_0 denote an arbitrary compact subdomain of Ω, that is, we suppose that the closure $\overline{\Omega}_0$ of Ω_0 is compact and lies entirely in Ω. The restriction of M to Ω_0 is a minimal surface M_0 of finite area. Consider an arbitrary smooth deformation M_t of M_0 with support lying inside Ω_0. We recall that the support of a deformation is the smallest closed subset of the surface outside which deformation does not occur.

Let $A(t) = \operatorname{vol}_2 M_t$ denote the area of the surface M_t. Then minimality of M means that $dA/dt|_{t=0} = 0$ for any such M_0 and M_t.

Clearly, any small deformation of M_0 specified by a graph is realized in the class of graphs. Each such deformation is described by the family of functions

$$z = f_t(x, y) = F(x, y, t), \qquad F(x, y, 0) = f(x, y), \qquad (x, y) \in \Omega_0,$$

where the functions f_t are equal to f outside the support K of the deformation F.

The area of the surface M_t given by the graph $z = f_t(x, y) = F(x, y, t)$ over the domain Ω_0 is calculated by means of the integral

$$A(t) = \int_{\Omega_0} \sqrt{1+F_x^2+F_y^2}\, dx\, dy.$$

Since M_0 is a minimal surface,

$$A'(0) = \int_{\Omega_0} \frac{d}{dt}\sqrt{1+F_x^2+F_y^2}|_{t=0}\, dx\, dy = 0.$$

Also,

$$\frac{d}{dt}\sqrt{1+F_x^2+F_y^2}|_{t=0} = \frac{f_xF_{xt}+f_yF_{yt}}{\sqrt{1+f_x^2+f_y^2}}, \quad \text{where } F_{*t} = \frac{d}{dt}F_*\bigg|_{t=0}.$$

We observe that

$$\frac{f_xF_{xt}}{\sqrt{1+f_x^2+f_y^2}} = \frac{\partial}{\partial x}\left(\frac{f_xF_t}{\sqrt{1+f_x^2+f_y^2}}\right) - \frac{\partial}{\partial x}\left(\frac{f_x}{\sqrt{1+f_x^2+f_y^2}}\right)F_t.$$

The integral over Ω_0 of the first term on the right-hand side of this equality is zero. In fact, when $t = 0$ the function F_t is zero outside K because there is no deformation outside the support. Extending the integrand by zero outside K to the whole plane $\mathbb{R}^2$ and considering the rectangle $Q =$

$[-a, a]\times[-b, b]$, which contains K, $Q\supset K$, we can go over to integration over Q:

$$\int_{\Omega_0}\frac{\partial}{\partial x}\left(\frac{f_xF_t}{\sqrt{1+f_x^2+f_y^2}}\right)=\int_K\frac{\partial}{\partial x}\left(\frac{f_xF_t}{\sqrt{1+f_x^2+f_y^2}}\right)$$
$$=\int_Q\frac{\partial}{\partial x}\left(\frac{f_xF_t}{\sqrt{1+f_x^2+f_y^2}}\right)=\int_{-b}^{b}\left[\int_{-a}^{a}\frac{\partial}{\partial x}\left(\frac{f_xF_t}{\sqrt{1+f_x^2+f_y^2}}\right)dx\right]dy$$
$$=\int_{-b}^{b}\frac{f_xF_t}{\sqrt{1+f_x^2+f_y^2}}\Bigg|_{x=-a}^{x=a}dy=0.$$

Similarly we consider the term $f_yF_{yt}/\sqrt{1+f_x^2+f_y^2}$. Thus,

$$A'(0)=-\int\left[\frac{\partial}{\partial x}\left(\frac{f_x}{\sqrt{1+f_x^2+f_y^2}}\right)+\frac{\partial}{\partial y}\left(\frac{f_y}{\sqrt{1+f_x^2+f_y^2}}\right)\right]F_t\,dx\,dy.$$

Since we can take for F_t any smooth function with support in Ω_0 (the support of the function is the smallest closed set outside which the function vanishes) and construct from it the corresponding deformation F, the expression in square brackets must vanish. (If it is not zero at a point (x_0, y_0), then it is not zero in a whole neighborhood of (x_0, y_0). It remains to construct a smooth nonnegative function with support in this neighborhood that does not vanish identically.)

The resulting condition is necessary and sufficient for the function f to specify a minimal surface M. Elementary calculation reduces the equation

$$\frac{\partial}{\partial x}\left(\frac{f_x}{\sqrt{1+f_x^2+f_y^2}}\right)+\frac{\partial}{\partial y}\left(\frac{f_y}{\sqrt{1+f_x^2+f_y^2}}\right)=0$$

to the minimal surface equation

$$(1+f_y^2)f_{xx}-2f_xf_yf_{xy}+(1+f_x^2)f_{yy}=0.$$

2. Bernstein's problem in $\mathbb{R}^3$. The following natural question arises: is there a function $z=f(x, y)$ that satisfies the minimal surface equation and is defined on the whole plane $\mathbb{R}^2$? Obviously, any linear function (specifying a plane) is a solution of this equation. Are there any nontrivial solutions? This a called *Bernstein's problem.* It turns out that in $\mathbb{R}^3$ there are no other complete graphs (graphs over the whole plane) that are minimal surfaces.

The solution of Bernstein's problem is based on two facts, which we prove in Chapter 3.

Let $z = f(x, y)$ be an arbitrary graph over the whole xy-plane that specifies a minimal surface M.

1. There is a change of coordinates $(x, y) \to (u, v)$, where (u, v) run through the whole plane $\mathbb{R}^2$, such that the metric on the surface M, written in the (u, v)-coordinates, has the form $ds^2 = \lambda(u, v)(du^2 + dv^2)$, where $\lambda(u, v)$ is a positive smooth function defined on the whole plane $\mathbb{R}^2_{(u,v)}$. This means that at each point of the graph M the velocity vectors of the coordinate curves $u = \text{const}$ and $v = \text{const}$ are orthogonal and have the same length. Such coordinates are called *isothermal* (for the details see §1 of Chapter 3).

2. Let us consider the Gaussian mapping ν of M, that is, with each point $(x, y, f(x, y))$ we associate the unit normal vector to M at this point that makes an acute angle with the z-axis. Since the unit vectors of all possible directions form the sphere S^2 given by the equation $x^2 + y^2 + z^2 = 1$, the mapping ν takes our minimal surface into the sphere S^2, $\nu: M \to S^2$. Clearly, the Gaussian image (the image of the Gaussian mapping) of any graph lies in the upper hemisphere.

On the sphere S^2 with the point P removed we can assign special coordinates, called *stereographic coordinates.* Let P be the North Pole of the sphere, that is, $P = (0, 0, 1)$. We map the sphere $S^2 \backslash P$ onto the xy-plane, associating with each point $Q \in S^2 \backslash P$ the point $\pi(Q)$ of the xy-plane at which the line PQ meets this plane (see Figure 28).

The mapping $\pi: S^2 \backslash P \to Oxy$ is called *stereographic projection.* Let (x, y) be the standard coordinates in the xy-plane. In the plane with coordinates (u, v) we form the corresponding complex coordinate $\xi = u + iv$, and in the xy-plane the coordinate $\eta = x + iy$.

The Gaussian mapping ν gives a mapping of the complex plane $\mathbb{C}_\xi \approx \mathbb{R}^2_{(u,v)}$ onto the complex plane $\mathbb{C}_\eta \approx \mathbb{R}^2 x, y$:

$$\mathbb{C}_\xi \approx \mathbb{R}^2_{(u,v)} \to M \xrightarrow{\nu} S^2 \backslash P \xrightarrow{\pi} \mathbb{R}^2 x, y \approx \mathbb{C}_\eta,$$

where $\mathbb{C}_\xi$ and $\mathbb{C}_\eta$ are the complex planes with coordinates ξ and η.

The main assertion of part 2 is that the mapping $\mathbb{C}_\xi \to \mathbb{C}_\eta$ defined in this way is holomorphic (complex-analytic). The proof of Bernstein's theorem follows immediately from this. In fact, the image of the Gaussian mapping ν of the surface M lies in the upper hemisphere. Therefore its image under the stereographic projection π is a bounded subset of $\mathbb{C}_\eta$. Thus, the Gaussian mapping ν specifies a bounded holomorphic function defined on the whole complex plane. By Liouville's theorem this function must be a constant. Consequently, the Gaussian mapping takes the whole surface M into one point. Therefore all the normals to the graph are parallel, and so the graph is a plane.

REMARK. From the proof it is clear that the main feature of part 1 is the assertion that the coordinates (u, v) run through the whole plane $\mathbb{R}^2$. Otherwise we could not apply Liouville's theorem.

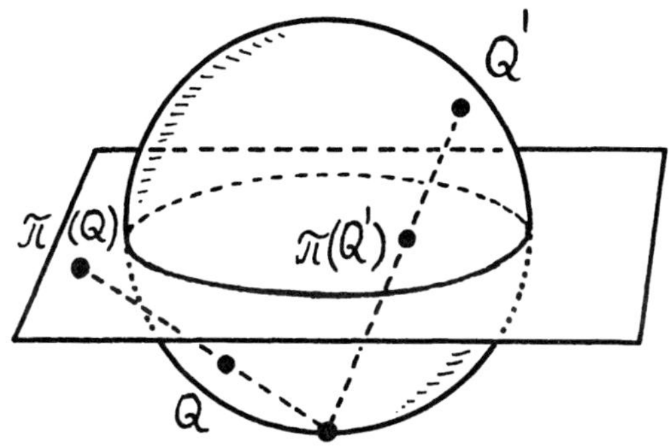

FIGURE 28

3. The Scherk surface and symmetry principle. One of the simplest types of functions of several variables is a function that can be represented as a sum of functions of each variable separately. In our case we consider a function $z = f(x, y) = \varphi(x) + \psi(y)$.

QUESTION. When does a function of this type specify a minimal surface?

This problem can be completely investigated. It is easy to see that in the given case the minimal surface equation has the form $(1 + \psi_y^2)\varphi_{xx} + (1 + \varphi_x^2)\psi_{yy} = 0$ and so it separates:

$$\frac{\psi_{yy}}{1+\psi_y^2} = -\frac{\varphi_{xx}}{1+\varphi_x^2} = a = \text{const}.$$

When $a = 0$ we obtain the equation of a plane. When $a \neq 0$ we have

$$\varphi(x) = \frac{1}{a}\ln[\cos(ax+b)] + c_1,$$

$$\psi(y) = -\frac{1}{a}\ln[\cos(ay+d)] + c_2,$$

where b, c_1, c_2, d are arbitrary constants. Thus,

$$z = f(x, y) = \frac{1}{a}\ln\frac{\cos(ax+b)}{\cos(ay+d)} + c.$$

Up to an isometry (a shift along the z-axis by c, and also along the x-axis and y-axis by b and d respectively) we have obtained a one-parameter family of minimal graphs, given by the functions $z = a^{-1}\ln(\cos ax/\cos ay)$, $a \neq 0$, each of which is defined over the "black squares of a chessboard", that is, where $\cos ax/\cos ay > 0$, for example, over $\{(x, y) | \, |ax| < \pi/2, |ay| < \pi/2\}$. These surfaces are called *Scherk surfaces*. As in the case of the catenoid and helicoid, we have a unique Scherk surface (up to isometry and homothety), given by the function $z = \ln(\cos x/\cos y)$.

Let us consider the fragment of the Scherk surface defined over the "black square" $|ax| < \pi/2$, $|ay| < \pi/2$. The intersection of this fragment with the xy-plane consists of two intervals $y = \pm x$, $|ax| < \pi/2$. This fragment goes into itself under rotation through an angle π about the z-axis and has a saddle form: when $|x| < |y|$ the graph is above the xy-plane, that is, $z > 0$, but when $|x| > |y|$ the graph is below the xy-plane.

As we approach a point of the boundary that is not a vertex of the black square, the z-coordinate of the graph tends to infinity. As we approach a vertex of the black square along different curves we can obtain in the limit any point of the vertical line passing through this vertex. Therefore the closure of the Scherk surface as a subset of $\mathbb{R}^3$ contains, apart from the graph itself, four vertical lines, passing through the vertices of the black square. These vertical lines are the boundary of a fragment of the Scherk surface. If we reflect this fragment about one of these vertical lines, then by the symmetry principle we obtain a minimal surface that is smooth along the vertical line. The resulting symmetrical fragment coincides with the fragment that is defined over a neighboring black square, and it is easy to see that it is smoothly combined into a single minimal surface with the fragments defined over the four adjacent black squares. Continuing this operation, we obtain the complete Scherk surface (see Figure 29).

This surface is a graph over the whole "black part" of the plane with the removal of a net consisting of lines parallel to the x-axis and the y-

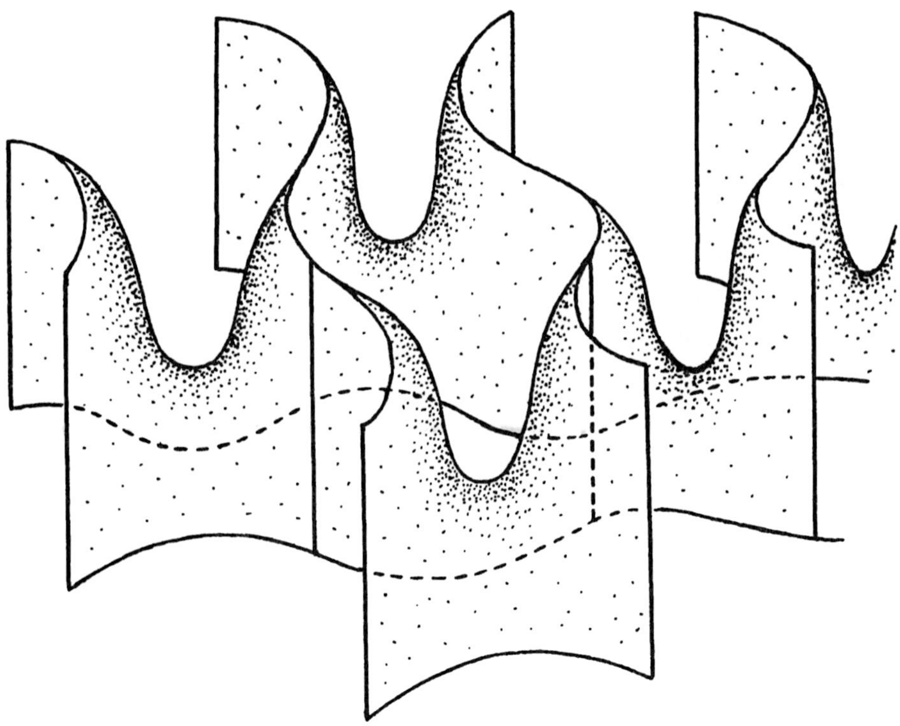

FIGURE 29

axis passing through the vertices of the lattice $\{x = (\pi/2 + \pi k)/a, y = (\pi/2+\pi l)/a\}$, $k, l \in \mathbb{Z}$. The fragments of the surface over adjacent squares are joined to each other by vertical lines passing through the vertices of the lattice.

§4. Periodic minimal surfaces

In §2 we stated the Schwarz-Riemann reflection principle 1, which enables us to extend minimal surfaces beyond rectilinear parts of the boundary. There is a second Schwarz reflection principle, which we now state.

THE SCHWARZ REFLECTION PRINCIPLE 2. Suppose that a smooth minimal surface M in $\mathbb{R}^3$ orthogonally approaches a plane Π, that is, the intersection of Π and the boundary ∂M of the surface M is a smooth regular curve γ along which M and Π are orthogonal. Then, if M^* denotes the image of M under reflection in Π, the union $M \cup M^*$ is a smooth minimal surface: the minimal surfaces M and M^* join smoothly along γ.

The situation in which a minimal surface approaches a plane at a right angle occurs naturally in problems known as problems with a partially free boundary. Let us illustrate this by an example. We assemble a configuration consisting of the surface P of a physical body (for example, a sheet of plexiglass) and a wire contour Γ attached to this surface. If we dip this configuration into a soap solution and lift it out, then a soap film remains on it (see Figure 30).

The boundary of this film consists of two parts: the wire contour Γ (the fixed part of the boundary) and a curve γ, the trace left by the soap film on the surface P (the "free" part of the boundary).

We have already met a similar situation when we spoke about an experimental "solution" of the planar Steiner problem (see §2 of Chapter 1).

Schwarz observed that in a problem with a partially free boundary the

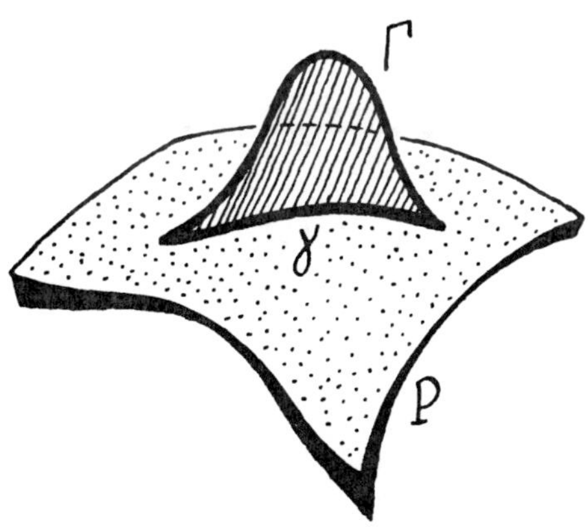

FIGURE 30

soap film spanning a configuration consisting of a set of surfaces and curves joining them (for example, $P \cup \Gamma$) approaches the surfaces occurring in this configuration at a right angle.

We now consider a connected configuration consisting of a set of flat surfaces $P_1, P_2, \dots, P_k$ and segments of lines $l_1, \dots, l_m$ (a so-called Schwarz chain (see Figure 31)).

Suppose that a soap film M is stretched on this configuration. Suppose that the surface M approaches (orthogonally) the planes $P_{i_1}, \dots, P_{i_r}$ and that $\gamma_{i_1}, \dots, \gamma_{i_r}$ are the traces left by this surface on the corresponding planes. Suppose that the boundary ∂M of M consists of the "traces" $\gamma_{i_1}, \dots, \gamma_{i_r}$ and the segments $l_{i_1}, \dots, l_{i_s}$ of the configuration. Then, in accordance with the reflection principles 1 and 2, we can extend M to a smooth minimal surface by reflecting M in the planes $P_{i_1}, \dots, P_{i_r}$ and the segments $l_{i_1}, \dots, l_{i_s}$. Obviously, the boundary of the resulting minimal surface again consists of segments of lines and "traces" lying in planes perpendicular to the extended surface along these "traces". We do not exclude the case when the extended surface has no boundary at all. If the surface has a boundary, then we can again extend the surface, and this extension process breaks off only when the boundary of the extended surface "disappears". Such a disappearance can happen, for example, when the configuration consists of one plane, and the film is noncompact and is a half-plane orthogonally approaching the plane of the configuration. If the initial film M is bounded, we can show that the extension process cannot break off after finitely many steps (since there are no closed, that is, compact without boundary, minimal surfaces in R^3; see §5). Let $\widetilde{M}$ denote the minimal surface obtained as a result of all possible extensions. Generally speaking, the surface $\widetilde{M}$ can have self-intersections. If the infinitely extended surface is embedded, that is, it does not have self-intersections and is a regular surface, then it is called the *periodic minimal surface* generated by M.

Back in 1867 Schwarz mentioned that the only minimal surface that spans

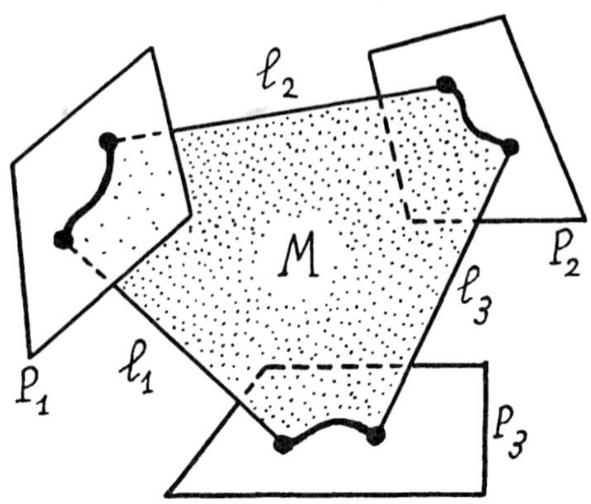

FIGURE 31

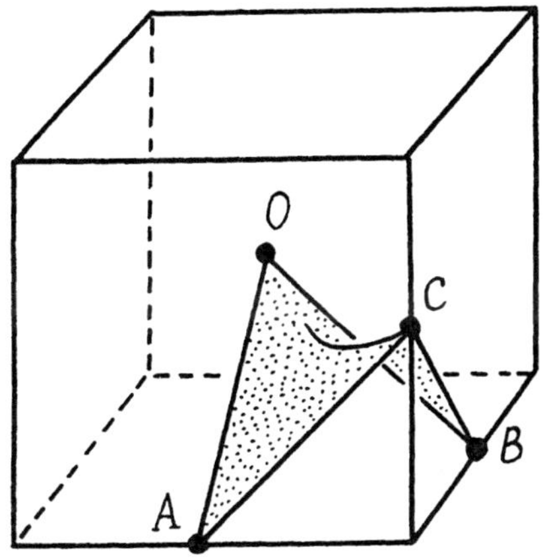

FIGURE 32

a spatial quadrilateral consisting of edges of a regular tetrahedron can be extended by means of reflections to a periodic minimal surface.

In fact, let us consider a cube and choose four points in it: the midpoints A, B, and C of the three edges meeting at one vertex and the center O of the cube. Clearly, these four points are the vertices of a regular tetrahedron. Let us construct the spatial quadrilateral $OABC$ (see Figure 32).

If we reflect $OABC$ about sides of it that do not lie in faces of the cube, that is, about the lines OA and OB, then it is easy to see that we again obtain a quadrilateral consisting of the vertex O and the midpoints of three edges meeting at a vertex of the cube. Repeating similar reflections for the resulting quadrilaterals, we obtain as a result six congruent spatial quadrilaterals, each of which consists of the center of the cube and the midpoints of three edges meeting at a vertex of the cube. Figures 33 and 34 show the trajectories of the points A and B, which together constitute the set of vertices of a regular plane hexagon, and also the trajectory of the point C under the reflections just described.

If we now stretch a minimal surface (soap film) on the contour $OABC$,

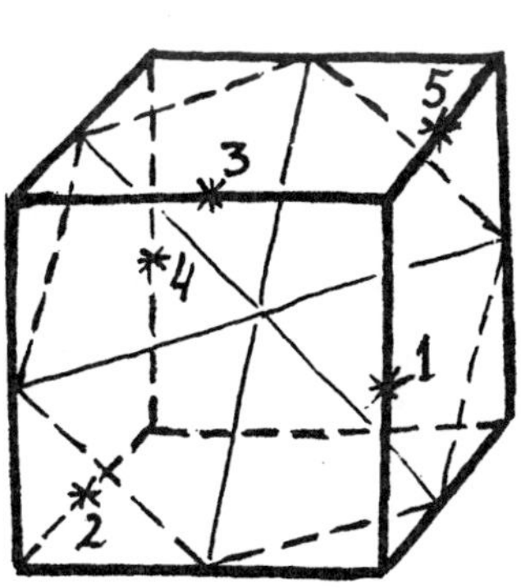

FIGURE 33A

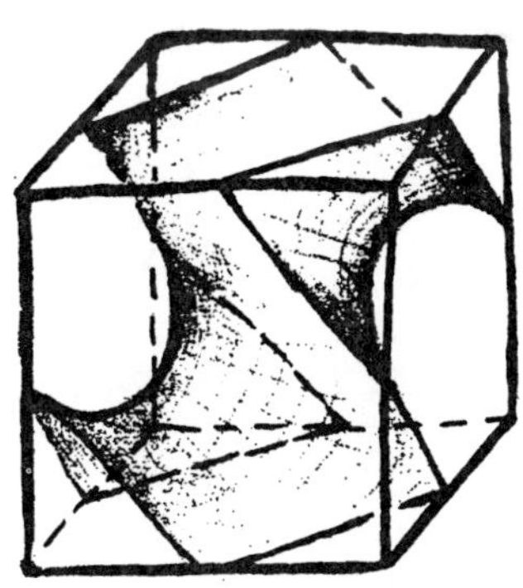

FIGURE 33B

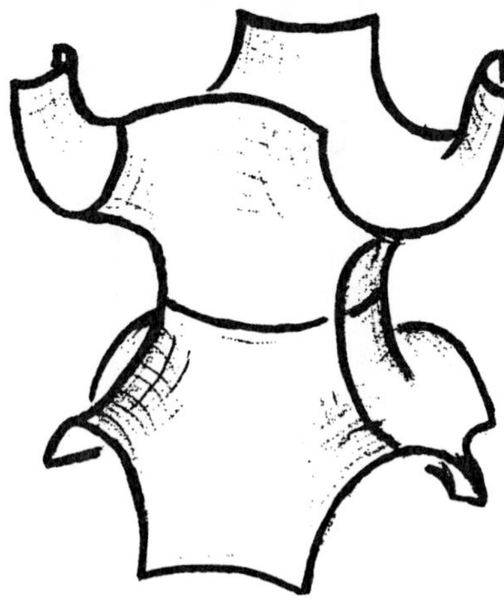

FIGURE 34

then as a result of the reflections just described we obtain, by the symmetry principle, a smooth minimal surface lying entirely in the cube and having as boundary a closed polygon consisting of twelve segments, each of which lies entirely in a face of the cube and joins the midpoints of the corresponding edges.

It is now not difficult to show that by reflections of a "cubic cell" about the edges of the polygon we can obtain a periodic surface (without self-intersections). This periodic minimal surface is called the *Schwarz-Riemann surface.*

There are many other periodic minimal surfaces, interesting examples of which were constructed by the American physicist and mathematician A. Schoen and the Finnish mathematician Neovius (see [3]). Figure 35 shows a fragment of the second Schwarz periodic surface constructed by Schoen.

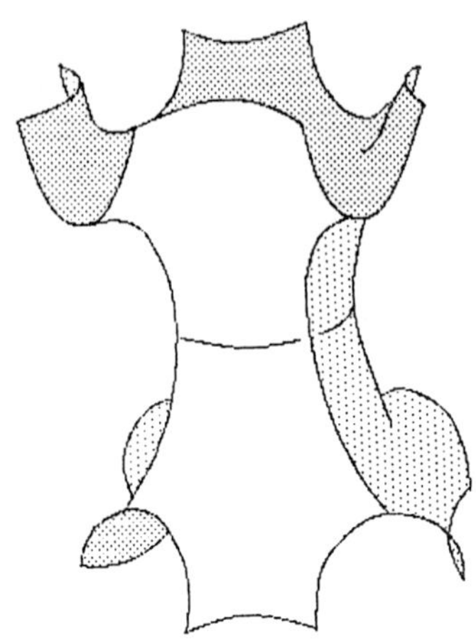

FIGURE 35

§5. Complete minimal surfaces

We have considered some basic examples of classical minimal surfaces in $\mathbb{R}^3$. Let us observe one general detail: these surfaces are either noncompact or compact, but then necessarily with a nonempty boundary. In fact, in $\mathbb{R}^3$ there are no closed (compact without boundary) minimal surfaces.

To prove this we need the so-called maximum principle. (In §2 we have already said that many properties of harmonic and analytic functions carry over to minimal surfaces. We now present a version of the maximum principle that is well known from the theory of harmonic and analytic functions.)

Let M_1 and M_2 be two minimal surfaces embedded in $\mathbb{R}^3$, touching each other at a point P that is internal for M_1 and M_2, and Π the common tangent plane to M_1 and M_2 at P. Locally (in a neighborhood of P) we specify the surface M_i by the graph $z = f_i(x, y)$, $i = 1, 2$, over Π, where (x, y) are coordinates in Π, the z-axis is perpendicular to Π, and $P = (0, 0, 0)$. We say that M_1 locally lies on one side of M_2 if in some neighborhood of P either $f_1(x, y) \geq f_2(x, y)$ or $f_1(x, y) \leq f_2(x, y)$.

THE MAXIMUM PRINCIPLE. Let M_1 and M_2 be two minimal surfaces embedded in $\mathbb{R}^3$, touching at a point P that is interior for M_1 and M_2. Suppose that M_1 locally (in a neighborhood of P) lies on one side of M_2. Then M_1 coincides with M_2 in some neighborhood of P.

We shall prove by contradiction that there is no closed minimal surface in $\mathbb{R}^3$. Let M be a closed minimal surface in $\mathbb{R}^3$. Then since M is bounded as a subset of $\mathbb{R}^3$ there is a plane that does not intersect M. We move this plane parallel to itself in the direction of M until it first touches M at some point P (we denote the tangent plane by Π). Clearly, P is an internal point of M and the plane Π (M has no boundary points because it is closed) and M lies on one side of Π. By the maximum principle M is flat in some neighborhood of P. In accordance with the uniqueness theorem (see §2) the connected component of M containing P is a flat closed surface, that is, an open-and-closed bounded subset of a plane. This contradicts the fact that a plane is connected. This completes the proof.

Thus, in $\mathbb{R}^3$ any minimal surface without boundary is noncompact. Among minimal surfaces without boundary an important class is formed by complete minimal surfaces. Such surfaces are maximal in some sense. As an illustration we consider a catenoid M given by $r = \cosh z$, $z \in \mathbb{R}$. If we restrict the domain of variation of z, say $z \in [a, b]$, or more generally take any proper subdomain of the catenoid, then the resulting minimal surface M_0 can be extended to a large connected minimal surface (for example, to the whole catenoid M). We note that if there is a nontrivial extension of some (not necessarily minimal) surface M without boundary to a large surface, then M, regarded as a subset of $\mathbb{R}^3$, is not closed. The catenoid M is a closed subset of $\mathbb{R}^3$, so it is impossible to extend it to a large surface. Applying the uniqueness theorem (see §2), we can conclude that any extension of M_0 to a large connected minimal surface is a subdomain of M. Thus, M is the largest minimal surface among all extensions of M_0 in the class of connected minimal surfaces. Moreover, the catenoid M determines all such extensions. Similar results also hold in the case when M is a helicoid, a Scherk surface, or a Schwarz-Riemann surface.

There is an alternative way of defining the completeness of a surface in terms of the completeness of the induced metric. In this direction there are two possibilities. Let M be an arbitrary connected embedded surface. On M we can define two distance functions, $|\ |$ and ρ. If A and B are arbitrary points on M, we put $|AB|$ equal to the Euclidean distance between A and B as points in $\mathbb{R}^3$. We call the function $|\ |$ the *extrinsic metric.* We define the function $\rho(A, B)$ as the greatest lower bound of the lengths of piecewise-smooth curves lying on M and joining A and B. We call the function ρ the *intrinsic metric.*

Thus, for an embedded surface M we have two metric spaces, $(M, |\ |)$ and (M, ρ).

We recall that a metric space is said to be *complete* if any fundamental sequence has a limit lying in this space.

ASSERTION 1. *Let M be an arbitrary surface embedded in $\mathbb{R}^3$. Then the metric space $(M, |\ |)$ is complete if and only if M is closed as a subset of $\mathbb{R}^3$.*

EXERCISE 1. Prove Assertion 1.

Thus, if M is a catenoid, a helicoid, a Scherk surface, or a Schwarz-Riemann surface, then the metric space $(M, |\ |)$ is complete.

We have been able to call a surface M complete if the metric space $(M, |\ |)$ is complete. However, in geometry there is another definition of completeness of a surface, imposing weaker restrictions, which nevertheless extends to a wider class of surfaces and turns out to be quite sufficient to obtain meaningful results.

DEFINITION. Let M be an arbitrary immersed connected surface, and ρ the intrinsic metric defined above. Then M is said to be *complete* if the metric space (M, ρ) is complete.

If M is an embedded connected surface, then it is easy to see that $\rho(A, B) \geq |AB|$ for any points A and B of M. Therefore the completeness of the metric space $(M, |\ |)$ implies the completeness of the metric space (M, ρ), that is, the completeness of M. The converse is false. Consider a spiral-shaped cylinder defined as follows: the directrix of M is a spiral γ lying in the xy-plane, which at one end tends asymptotically to a circle S^1 and at the other end to infinity; the generator of M is a line perpendicular to the xy-plane. Clearly, M is not closed as a subset of $\mathbb{R}^3$—the cylinder over S^1 lies in the closure of M and does not belong to M. Therefore the metric space $(M, |\ |)$ is not complete. Nevertheless the surface M is complete (verify this).

What does completeness of the metric space (M, ρ) imply? We note that a metric on an arbitrary set always determines a class of bounded subsets. We recall that a subset Y of a metric space (X, ρ) is said to be *bounded* if the distances between all possible points A and B of Y are bounded above: $\rho(A, B) \leq d$ for some $d \geq 0$. If M is a surface embedded in $\mathbb{R}^3$, then the metrics $|\ |$ and ρ defined above have different classes of bounded sets, for example in the case of a spiral-shaped cylinder.

Let M be a surface immersed in $\mathbb{R}^3$, and ρ the intrinsic metric.

ASSERTION 2. *The metric space (M, ρ) is complete if and only if any closed bounded (in the metric ρ) subset of M is a closed subset of $\mathbb{R}^3$.*

EXERCISE 2. Prove Assertion 2, using the following lemma.

LEMMA (see [7]). *An immersed surface M is complete if and only if any closed bounded (in the metric ρ) subset of M is compact.*

Thus, completeness of a surface M implies that to a bounded (in the intrinsic metric ρ) subset of M we cannot add any point of $\mathbb{R}^3$ that is the limit of points of M but does not lie on M.

For deeper results that follow from the completeness of a surface M, see [7]. In particular, the following assertion is true.

ASSERTION 3 (see [7]). *Let M be a complete connected surface without boundary immersed in $\mathbb{R}^3$. Then M cannot be extended: there is no connected surface M_1 immersed in $\mathbb{R}^3$ that contains M as a proper (not coinciding with M_1) subdomain.*

REMARK. It should not be thought that any incomplete immersed surface is extendable. Let M' be the xy-plane with the origin removed, and let (r, φ) be polar coordinates on M', where $r \in \mathbb{R}_+$ and $\varphi \in [0, 2\pi]$. If we allow the angle φ to vary from $-\infty$ to $+\infty$, we obtain an immersed surface M—an infinite-sheeted winding of the plane $\mathbb{R}_+ \times \mathbb{R} \approx \mathbb{R}^2$ onto the plane without the origin (compare with the case of a catenoid in §2). Clearly, the immersed surface is minimal and not complete.

EXERCISE 3. Prove that the surface M cannot be extended.

CHAPTER 3

General Properties of Minimal Surfaces in $\mathbb{R}^3$

We go over to a more detailed description of minimal surfaces in $\mathbb{R}^3$. In Chapter 2 we demonstrated how, by means of the symmetry principle, the uniqueness theorem, and the maximum principle, we can obtain various nontrivial results (Catalan's theorem, and the nonexistence of a closed minimal surface in $\mathbb{R}^3$). In this chapter we consider these and many other principles in more detail and give proofs of them.

In §1 we prove that on a minimal surface there are isothermal (conformal) coordinates that induce a so-called conformal structure. In these coordinates many formulas are significantly simplified.

In §2 we consider the Gaussian mapping of surfaces and prove that for minimal surfaces the Gaussian mapping is anticonformal. We have used this property in the solution of Bernstein's problem. It also plays an important role in the investigation of indices of minimal surfaces.

In §3, relying on the results of the first two sections, we construct the Weierstrass representation, which describes the local structure of all minimal surfaces in $\mathbb{R}^3$ by two complex-valued functions.

In §4 we define the global Weierstrass representation for oriented and nonoriented surfaces that are minimal.

In §§5 and 6 we talk about the investigation of complete minimal surfaces that have finite total curvature.

The concluding section (§7) is devoted to the investigation of indices of surfaces in $\mathbb{R}^3$.

§1. Isothermal coordinates

Let M be a regular surface in $\mathbb{R}^3$, and let (u, v) be local coordinates on M.

DEFINITION. Local coordinates (u, v) on a surface M are called *isothermal* (conformal) if the metric ds^2 on M induced from $\mathbb{R}^3$, described in the (u, v)-coordinates, is as follows: $ds^2 = \lambda(u, v)(du^2+dv^2)$, where $\lambda(u, v)$ is a positive function, called the *conformal factor*. In other words, in isothermal coordinates the tangential coordinate vectors at each point are perpendicular and have the same length.

Theorem (existence of isothermal coordinates; see [8]). *Let M be an arbitrary regular (of class C^2) surface in $\mathbb{R}^3$. Then for any point P of M there is a neighborhood U, $P \in U$, and local coordinates (u, v) in this neighborhood U such that the induced metric ds^2 on M, written in the (u, v)-coordinates, is as follows:* $ds^2 = \lambda(u, v)(du^2 + dv^2)$.

Remark. For a real-analytic surface M the construction of isothermal coordinates is significantly simplified: it actually reduces to the solution of differential equation [8].

We give a proof of this theorem for minimal surfaces due to Osserman [9]. It turns out that in this case we obtain explicit formulas for the isothermal coordinates. We choose a Cartesian system of coordinates x, y, z with origin at P so that the xy-plane coincides with the tangent plane T_PM to M at P. In this coordinate system the surface M in a neighborhood of P can be given by a graph $x = x$, $y = y$, $z = f(x, y)$. All such functions f are described by the minimal surface equation (see §1 of Chapter 2):

$$(1 + f_y^2)f_{xx} - 2f_xf_yf_{xy} + (1 + f_x^2)f_{yy} = 0.$$

Let us form three functions:

$$\frac{1 + f_x^2}{\sqrt{1 + f_x^2 + f_y^2}}, \quad \frac{f_xf_y}{\sqrt{1 + f_x^2 + f_y^2}}, \quad \frac{1 + f_y^2}{\sqrt{1 + f_x^2 + f_y^2}}.$$

Henceforth for convenience we put $p = f_x$, $q = f_y$, and $w = \sqrt{1 + f_x^2 + f_y^2}$. We recall that $\sqrt{1 + f_x^2 + f_y^2}\,dx\,dy$ is the element of area of the surface M.

It turns out that from the minimal surface equation we have the following relations:

$$\left(\frac{1+p^2}{w}\right)_y = \left(\frac{pq}{w}\right)_x, \quad \left(\frac{1+q^2}{w}\right)_x = \left(\frac{pq}{w}\right)_y.$$

In fact, direct calculation shows that

$$\left(\frac{1+f_x^2}{w}\right)_y - \left(\frac{f_xf_y}{w}\right)_x = -\frac{f_y}{w}[(1 + f_y^2)f_{xx} - 2f_xf_yf_{xy} + (1 + f_x^2)f_{yy}],$$

$$\left(\frac{1+f_y^2}{w}\right)_x - \left(\frac{f_xf_y}{w}\right)_y = -\frac{f_x}{w}[(1 + f_y^2)f_{xx} - 2f_xf_yf_{xy} + (1 + f_x^2)f_{yy}].$$

Therefore in any simply-connected neighborhood of the origin in the plane T_PM there are functions $F(x, y)$ and $G(x, y)$ such that

$$\frac{\partial F}{\partial x} = \frac{1+p^2}{w}, \qquad \frac{\partial F}{\partial y} = \frac{pq}{w},$$

$$\frac{\partial G}{\partial x} = \frac{pq}{w}, \qquad \frac{\partial G}{\partial y} = \frac{1+q^2}{w}.$$

Let us specify a map Φ: $(x, y) \to (u, v)$ by putting

$$u = x + F(x, y), \qquad v = y + G(x, y).$$

Then (u, v) are isothermal coordinates. In fact, the Jacobian J of the map Φ is equal to $\partial(u, v)/\partial(x, y) = (1+w)^2/w$, so $J > 0$, and so Φ is a local diffeomorphism. Thus, in some neighborhood of the origin the map Φ has a differential inverse Φ^{-1} with Jacobi matrix equal to

$$\begin{pmatrix} x_u & x_v \\ y_u & y_v \end{pmatrix} = \begin{pmatrix} 1+w+q^2 & -pq \\ -pq & 1+w+p^2 \end{pmatrix} \frac{1}{(1+w)^2}.$$

In the (u, v)-coordinate the surface M is given by the radius vector

$$r(u, v) = (x(u, v), y(u, v), f(x(u, v), y(u, v))).$$

Direct calculation shows that

$$|r_u|^2 = |r_v|^2 = \frac{w^2}{(1+w)^2}, \qquad \langle r_u, r_v\rangle = 0,$$

where $\langle\ ,\ \rangle$ is the standard Euclidean scalar product in $\mathbb{R}^3$.

REMARK. In §3 of Chapter 2 we talked about Bernstein's problem. If the graph $z = f(x, y)$ defining a minimal surface is defined on the whole xy-plane, we can extend the functions $F(x, y)$ and $G(x, y)$ to the whole plane, and hence define a mapping on the whole plane. As we remarked above, this mapping is a local diffeomorphism. We now show that Φ is actually a diffeomorphism of the xy-plane onto the whole uv-plane. For this it is sufficient to show that Φ does not decrease distances between points. In fact, if this is so, then any two distinct points go into a pair of noncoincident points. Therefore Φ is bijective and is therefore a diffeomorphism with image $\operatorname{Im}\Phi$. Suppose that $\operatorname{Im}\Phi$ does not coincide with the whole uv-plane. Let $Q = \Phi(P)$, and let Q' be a boundary point of the set $\operatorname{Im}\Phi$. Since Φ is a diffeomorphism, Q' does not lie in the image of Φ. Suppose that a sequence Q_k of points of $\operatorname{Im}\Phi$ converges to Q' and that $P_k = \Phi^{-1}(Q_k)$. Obviously, the sequence P_k does not converge to any point P' of the xy-plane. Therefore the distance $|PP_k|$ tends to infinity as $k \to \infty$, and so

$$|QQ'| = \lim_{k\to\infty} |QQ_k| \geq \lim_{k\to\infty} |PP_k| = \infty.$$

We have obtained a contradiction. Thus, it remains to show that Φ does not decrease distances.

We observe that Φ is the sum of the two mappings, the identity id and the mapping ψ: $(x, y) \mapsto (F(x, y), G(x, y))$. We now show that ψ does not strongly twist the plane. More precisely, let P and P' be two distinct points of the xy-plane (which we shall identify with the points of the uv-plane having the same coordinates), and Q and Q' their images under the map ψ.

Lemma (see also [9]). *The angle between the vectors QQ' and PP' is acute.*

Proof. Consider the segment joining P and P', defined as follows: $(1-t)P + tP'$, $t \in [0, 1]$. Let $G(t)$ be equal to the scalar product of the radius vector $\psi((1-t)P + tP')$ and the vector

$$P' - P\colon\ G(t) = \langle \psi((1-t)P + tP'), P' - P\rangle.$$

Then $\dot{G}(t) = \langle H(P'-P), P'-P\rangle$, where

$$H = \begin{pmatrix} \partial F/\partial x & \partial F/\partial y \\ \partial G/\partial x & \partial G/\partial y \end{pmatrix} = \begin{pmatrix} (1+p^2)/w & pq/w \\ pq/w & (1+q^2)/w \end{pmatrix}$$

is the Jacobi matrix of ψ. The matrix H is positive definite. In fact, the principal minors $(1+p^2)/w$ and $\det H = 1$ are positive. Consequently, $\dot{G}(t) > 0$, so $G(1) > G(0)$. This inequality is equivalent to $\langle Q' - Q, P' - P\rangle > 0$. This completes the proof of the lemma.

It is easy to see that the sum of the identity transformation and a mapping that does not strongly twist the plane does not decrease distances. The proof of this assertion is illustrated in Figure 36; a strict proof is left to the reader as an exercise.

Let us give some examples of global isothermal coordinates. Let (r, φ, z) be cylindrical coordinates in $\mathbb{R}^3$. Then on the catenoid $r = \cosh z$ the coordinates (φ, z) are isothermal: $ds^2 = \cosh^2 z(d\varphi^2 + dz^2)$. On the helicoid, given in Cartesian coordinates (x, y, z) as

$$x = \sinh u \cos v, \qquad y = \sinh u \sin v, \qquad z = v;$$

the coordinates (u, v) are also isothermal: $ds^2 = \cosh^2 u(du^2 + dv^2)$. Below we construct the Weierstrass representation, which also gives minimal surfaces in isothermal coordinates.

A very useful consequence of the theorem on the existence of isothermal coordinates is the possibility of introducing a conformal structure on a surface.

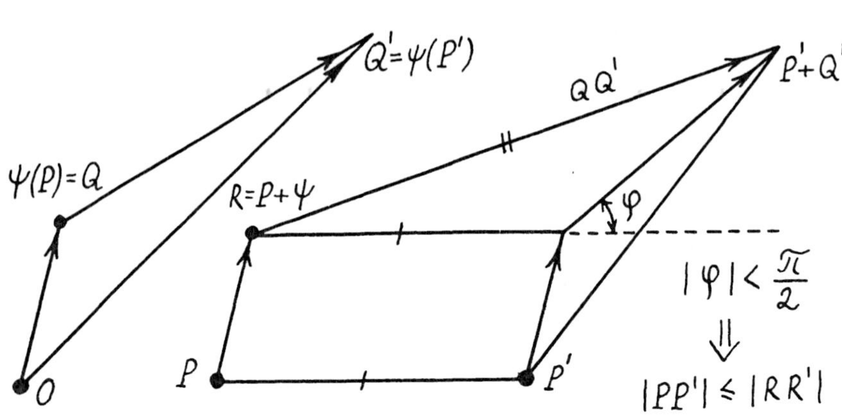

Figure 36

Lemma. *Let (u, v) and (u', v') be isothermal coordinates defined in some domain U of a surface M. Then a change of coordinates $(u, v) \to (u', v')$ that preserves the orientation is a conformal transformation.*

We recall that a conformal transformation is a diffeomorphism $f\colon V \to W$ of a planar domain $V \subset \mathbb{R}^2$ onto a planar domain $W \subset \mathbb{R}^2$ such that the differential df of the map f is a composition of a rotation and an expansion (with respect to the metrics in V and W).

To prove the lemma, we need to show that a linear transformation that preserves the conformal form of the scalar product is a composition of an orthogonal transformation and an expansion (verify this).

We now represent each point of a domain $V \subset \mathbb{R}^2$ with coordinates (u, v) as a point in the complex plane $\mathbb{C} \approx \mathbb{R}^2$ with complex coordinate $z = u + iv$. Each tangent vector $ae_1 + be_2$, where e_1 and e_2 are the tangent velocity vectors of the coordinate curves $v = \text{const}$ and $u = \text{const}$ respectively, can now be written as $(a + ib)e$, where $e = e_1$, $ie = e_2$. We proceed in the same way with a domain $W \subset \mathbb{R}^2$ with coordinates (u', v').

Then $f\colon V \to W$ is a complex-valued function on V. How do we describe the linear mapping df in complex form? Any linear mapping $L\colon (a, b) \to (a', b')$ can be represented by a matrix

$$\begin{pmatrix} \alpha + \beta & \gamma + \delta \\ -(\gamma - \delta) & (\alpha - \beta) \end{pmatrix}$$

that is,

$$(a', b') = \begin{pmatrix} \alpha & \gamma \\ -\gamma & \alpha \end{pmatrix} \begin{pmatrix} a \\ b \end{pmatrix} + \begin{pmatrix} \beta & \delta \\ -\delta & \beta \end{pmatrix} \begin{pmatrix} a \\ -b \end{pmatrix}.$$

If we introduce complex coordinates $w = a + ib$ and $w' = a' + ib'$, then our transformation can be written as $w' = (\alpha + i\gamma)w + (\beta + i\delta)\overline{w}$. It is easy to show that if L is the composition of an orthogonal transformation and an expansion, then it can be written either as $w' = (\alpha + i\gamma)w$ (L preserves the orientation) or as $w' = (\beta + i\delta)\overline{w}$ (L reverses the orientation). Therefore for a conformal mapping $f\colon V \to W$ the differential df at each point can be represented by a complex number which acts on the complex coordinate $w = a + ib$ of the vector $ae_1 + be_2 = (a + ib)e$ by multiplication. The second component in the expansion of df, that is, $(\beta + i\delta)\overline{w}$, is equal to zero for any $\overline{w}$.

If $f = (\varphi(u, v), \psi(u, v)) = \varphi + i\psi$, then

$$df = \begin{pmatrix} \varphi_u & \varphi_v \\ \psi_u & \psi_v \end{pmatrix} = \begin{pmatrix} \alpha + \beta & \gamma + \delta \\ -(\gamma - \delta) & \alpha - \beta \end{pmatrix},$$

so

$$df(a, b) = \frac{1}{2} \begin{pmatrix} \varphi_u + \psi_v & \varphi_v - \psi_u \\ -(\varphi_v - \psi_u) & \varphi_u + \psi_v \end{pmatrix} \begin{pmatrix} a \\ b \end{pmatrix} + \frac{1}{2} \begin{pmatrix} \varphi_u - \psi_v & \varphi_v + \psi_u \\ -(\varphi_v + \psi_u) & \varphi_u - \psi_v \end{pmatrix} \begin{pmatrix} a \\ -b \end{pmatrix}.$$

Therefore the condition that f is conformal is the condition that the second matrix is zero, that is, $\varphi_u = \psi_v$, $\varphi_v = -\psi_u$, and this is the Cauchy-Riemann holomorphy condition that the complex-valued function $f = f(z)$ is complex-analytic. Moreover, the complex number $\frac{1}{2}[(\varphi_u+\psi_v)+i(\varphi_v-\psi_u)]$ is equal to the complex derivative of f with respect to z, that is, $df/dz = f_z$.

CONCLUSION. On an oriented surface M all possible isothermal coordinates such that the transition functions preserve orientation specify a complex structure. Each pair of functions (u, v) giving isothermal coordinates is replaced by the complex coordinate $z = u + iv$, and the change of coordinate functions are complex-analytic. In this notation, to differentials of change of coordinate mappings there correspond the usual complex derivatives of the relevant holomorphic functions, which act on the complex coordinates of the tangent vectors by multiplication. Surfaces with a complex structure are called *Riemann surfaces.*

§2. Harmonicity and conformality

The theory of two-dimensional minimal surfaces in $\mathbb{R}^3$ relies substantially on the possibility of introducing isothermal coordinates in a neighborhood of any point of the surface. Let us consider some examples.

1. Suppose that a regular surface M in $\mathbb{R}^3$ is given locally by a radius vector $r(u, v) = (x(u, v), y(u, v), z(u, v))$, where (u, v) are isothermal coordinates on M. Let $\Delta = \partial^2/\partial u^2+\partial^2/\partial v^2$ denote the standard Euclidean Laplacian.

PROPOSITION 1. *The radius vector* $\Delta r = (\Delta x, \Delta y, \Delta z)$ *is perpendicular to the surface* M.

PROOF. Since (u, v) are isothermal coordinates on M, we have $\langle r_u, r_u\rangle = \langle r_v, r_v\rangle$ and $\langle r_u, r_v\rangle = 0$. Differentiating the first equality with respect to u and the second with respect to v, we obtain

$$\langle r_{uu}, r_u\rangle = \langle r_{uv}, r_v\rangle, \qquad \langle r_{vv}, r_u\rangle = -\langle r_{uv}, r_v\rangle.$$

Therefore $\langle \Delta r, r_v\rangle = 0$. Similarly, $\langle \Delta r, r_u\rangle = 0$.

PROPOSITION 2. *The radius vector* Δr *is equal to* $2\lambda H n$, *where* H *is the mean curvature of* M, n *is the unit normal to* M, *and* λ *is the conformal factor of the induced metric* ds^2 *on* M, *where* $ds^2 = \lambda(u, v)(du^2 + dv^2)$. *If* M *is a minimal surface, then in isothermal coordinates* (u, v) *the radius vector* $r(u, v)$ *that locally describes the surface* M *satisfies the equation* $\Delta r = 0$.

DEFINITION. The radius vector $r(u, v)$ of a surface M specified in the (u, v)-coordinates that satisfies the condition $\Delta r = 0$, where $\Delta = \partial^2/\partial u^2 + \partial^2/\partial v^2$, is called a *harmonic* radius vector.

COROLLARY 1. *In isothermal coordinates the minimality of a surface is equivalent to the harmonicity of the radius vector specifying it.*

REMARK. A generalization of a harmonic radius vector is the concept of a harmonic mapping. The theory of harmonic mappings is a very elegant and well developed branch of mathematics. A reader who is interested in it can turn to [39] and [40].

Before proving Proposition 2, we give a more formal construction that enables us to define the principal curvatures and the mean curvature and to give a direct proof of Euler's formula and Meusnier's theorem, described in Chapter 1. Suppose that a surface M in $\mathbb{R}^3$ is specified by a radius vector $r(u, v)$, where (u, v) are local coordinates. Let n be the unit normal to M. We define a quadratic form Q (called the *second fundamental form*) on the tangent vectors. If v is a tangent vector to M at a point $P \in M$, and $\gamma(t)$ is a curve on M with tangent vector v, where $\gamma(0) = P$ and $\dot{\gamma}(0) = v$, we put $Q(v) = \langle \ddot{\gamma}(0), n \rangle$. Thus, $Q(v)$ is the normal component of the acceleration vector of the curve γ. If $\gamma(t)$ is given in the (u, v)-coordinates as $(u(t), v(t))$, then $\ddot{\gamma} = r_{uu}\dot{u}^2 + 2r_{uv}\dot{u}\dot{v} + r_{vv}\dot{v}^2 + (r_u\ddot{u} + r_v\ddot{v})$. The expression in parentheses is the tangent vector to M. Therefore

$$Q(v) = \langle \ddot{\gamma}, n \rangle = b_{11}\dot{u}^2 + 2b_{12}\dot{u}\dot{v} + b_{22}\dot{v}^2,$$

where

$$b_{11} = \langle r_{uu}, n \rangle, \quad b_{21} = b_{12} = \langle r_{uv}, n \rangle, \quad b_{22} = \langle r_{vv}, n \rangle$$

and $(\dot{u}, \dot{v})$ are the coordinates of the tangent vector $v = \dot{\gamma}(0) = r_u\dot{u} + r_v\dot{v}$. We have obtained a quadratic form in the coordinates of the tangent vector (this implies that $Q(v)$ does not depend on the curve γ on condition that $P = \gamma(0)$ and $v = \dot{\gamma}(0)$). Therefore, if we take the normal section of M in the direction v, the curvature $\kappa(v)$ of this normal section will be $\kappa(v) = Q(v)/|v|^2$. Thus, the curvature $\kappa(v)$ is equal to $Q(v)$ if the vector v describes the unit circle in T_PM.

We can now define the principal curvatures and principal directions at a point $P \in M$. Let us choose an orthonormal basis in T_PM. Suppose that the second fundamental form $Q(v)$ is specified in it by the matrix $\overline{Q}$. Then the principal curvatures, that is, the maximum and minimum values of $\kappa(v)$ on the unit circle, will be equal to the eigenvalues of the matrix $\overline{Q}$, and the principal directions are the directions specified by the eigenvectors of $\overline{Q}$. The equation for the eigenvalues is $\det(\overline{Q} - \lambda E) = 0$, where E is the unit matrix; this is equivalent to the equation $\det[A^{\mathrm{T}}(\overline{Q} - \lambda E)A] = 0$, where A is any nonsingular matrix. If for A we take the matrix of transition from the orthonormal basis to the original basis, then the equation for the eigenvalues can be rewritten as $\det(Q - \lambda G) = 0$, where $G = A^{\mathrm{T}}A$ is the matrix of the first fundamental form (the metric). An eigenvector v of $\overline{Q}$, that is, a vector satisfying the equation $(\overline{Q} - \lambda E)v = 0$, can be rewritten in the original basis as $\overline{v} = A^{-1}v$, that is, it satisfies the equation

$$A^{\mathrm{T}}(\overline{Q} - \lambda E)AA^{-1}v = (Q - \lambda G)\overline{v} = 0.$$

Thus, the principal curvatures are equal to the invariants of a pair of quadratic forms, the first and second fundamental forms, and are equal to the eigenvalues of the matrix QG^{-1}, and the principal directions are given by the directions of the eigenvectors of this matrix.

The mean curvature H is equal to half the sum of the principal curvatures, that is, to half the sum of the eigenvalues of the matrix $\overline{Q}$ (in the orthogonal basis) or of the matrix QG^{-1} (in any basis). This means that $H = \frac{1}{2}\operatorname{tr} QG^{-1}$, where tr denotes the trace of a matrix. The Gaussian curvature K, the product of the principal curvatures, is equal to $K = \det QG^{-1} = \det Q/\det G$. Since the eigenvectors are perpendicular for different eigenvalues, the principal directions are also perpendicular.

REMARK. Euler's formula (Chapter 1) can now be obtained in an elementary way. It is sufficient to choose an orthonormal basis from the eigenvectors of $\overline{Q}$ and observe that in this basis $\overline{Q} = \left(\begin{smallmatrix}\lambda_1 & 0\\ 0 & \lambda_2\end{smallmatrix}\right)$, where λ_1 and λ_2 are the principal curvatures. If we take the unit tangent vector $v = (\cos\varphi, \sin\varphi)$, that is, the vector that makes an angle φ with a principal direction, then

$$Q(v) = \lambda_1\cos^2\varphi + \lambda_2\sin^2\varphi = \kappa(\varphi) = \kappa(v).$$

It is also easy to obtain a proof of Meusnier's theorem. It is sufficient to introduce the natural parameter t on the plane section $\gamma = \Pi_1 \cap M$, that is, to force the point to move along γ with unit speed.

It is easy to show that $\ddot{\gamma}$ is perpendicular to $\dot{\gamma}$ (we have $\langle\dot{\gamma}, \dot{\gamma}\rangle = 1$, and it remains to differentiate with respect to t). Therefore, $\ddot{\gamma} = \kappa n_1$ is the curvature vector of γ, where n_1 is the unit normal to $\dot{\gamma}$ in the plane Π_1. We have $Q(\dot{\gamma}) = \langle\ddot{\gamma}, n\rangle = \kappa\langle n_1, n\rangle = \kappa\cos\psi$, where ψ is the angle between n_1 and the normal n to M.

PROOF OF PROPOSITION 2. Let (u, v) be isothermal coordinates on an (arbitrary) surface M. Then

$$\langle\Delta r, n\rangle = \langle r_{uu}, n\rangle + \langle r_{vv}, n\rangle = b_{11} + b_{22}.$$

Since the matrix of the metric $G = \left(\begin{smallmatrix}\lambda & 0\\ 0 & \lambda\end{smallmatrix}\right)$ is diagonal, we have $H = \frac{1}{2}\operatorname{tr} QG^{-1} = (b_{11} + b_{22})/2\lambda$. Therefore $\langle\Delta r, n\rangle = 2\lambda H$.

The fact that the radius vector of a minimal surface in isothermal coordinates is harmonic has many profound consequences.

COROLLARY 2. *Let M be a connected compact smooth minimal surface in $\mathbb{R}^3$ with boundary ∂M. Then M lies in the convex hull $C(\partial M)$ of its boundary. In particular, any soap film spanning a wire contour lies in the convex hull of this contour.*

We recall that the convex hull of an arbitrary subset of $\mathbb{R}^3$ is the intersection of all closed half-spaces in $\mathbb{R}^3$ that contain the given subset completely.

PROOF. Suppose that M does not lie in the convex hull of its boundary ∂M. This means that there is a half-space $\{f \leq 0\}$ for some $f = \alpha x + \beta y + \gamma z + \delta$ such that ∂M lies in this half-space, but some point of M does not lie in it. Since M is compact, the function f, restricted to M, takes its maximum value at some point $P \in M$, and $f(P) > 0$. Therefore P is an interior point. If we introduce isothermal coordinates in a neighborhood of P, then the restriction of f to M in these coordinates is a harmonic function (since x, y, z are harmonic) that takes its maximum value at an interior point. Consequently, $f|_M = \text{const}$ in this neighborhood, that is, M is flat in this neighborhood. By the uniqueness theorem the minimal surface M is entirely flat. Now it is easy to obtain a contradiction.

COROLLARY 3 (a proof of the uniqueness theorem; see §2 of Chapter 2). *Let M_1 and M_2 be the two connected minimal surfaces described in the uniqueness theorem. Suppose that the uniqueness theorem is false. This implies that there is a point $P \in M_1 \cap M_2$ that is a boundary point of a maximal domain lying in $M_1 \cap M_2$ (which exists by the condition of the theorem). Consider the tangent plane Π to M_1 and M_2 at P and introduce coordinates (x, y, z) in a neighborhood of P, where (x, y) are Cartesian coordinates in Π, the z-axis is perpendicular to Π, and $P = (0, 0, 0)$. Then in a neighborhood of P the surfaces M_i are given by graphs $z = f_i(x, y)$, $i = 1, 2$, where $f_1 \not\equiv f_2$ in any neighborhood of P, and there is an open set $U \subset \Pi$ such that P lies on the boundary of U and $f_1 \equiv f_2$ in U.*

Carrying out the construction used in the proof of the theorem on the existence of isothermal coordinates, in some connected neighborhood of $P \in \Pi$ we construct isothermal coordinates (u, v) for each of the surfaces M_1 and M_2. The surface M_i is given locally by the radius vector $r_i(u, v)$, and in a domain $U \subset M_1 \cap M_2$ the two minimal surfaces are parametrized in the same way: $r_1(u, v) = r_2(u, v)$, $(u, v) \in U \subset \Pi$.

Since $x_i(u, v)$, $y_i(u, v)$, and $z_i(u, v)$ are harmonic functions, and

$$x_1(u, v) = x_2(u, v), \quad y_1(u, v) = y_2(u, v), \quad z_1(u, v) = z_2(u, v)$$

in an open set $U \subset \Pi$, they coincide in a whole neighborhood of P, that is, $r_1(u, v) = r_2(u, v)$, which contradicts the choice of the point P.

In §1 we showed that isothermal coordinates generate a complex structure on an oriented surface. On the other hand, in isothermal coordinates the radius vector that specifies a minimal surface is harmonic. As we know, harmonic functions are the real (or imaginary) parts of complex-analytic functions. From these two remarks it follows that an oriented minimal surface in $\mathbb{R}^3$ can be regarded as the real part of a holomorphic curve in $\mathbb{C}^3$; this, for example, gives the possibility of constructing an associated family of minimal surfaces (see the example of an associated family for the catenoid and helicoid in §2 of Chapter 2). We consider this situation in detail, first recalling some facts from complex differential geometry.

Let f be a complex-valued ($\mathbb{C}$-valued) function defined in a planar domain $U \subset \mathbb{R}^2 \approx \mathbb{C}$ with coordinates (u, v) and let $z = u + iv$ be the corresponding complex coordinate. Consider the complex differential operators

$$\frac{\partial}{\partial z} = \frac{1}{2}\left(\frac{\partial}{\partial u} - i\frac{\partial}{\partial v}\right) \quad \text{and} \quad \frac{\partial}{\partial \overline{z}} = \frac{1}{2}\left(\frac{\partial}{\partial u} + i\frac{\partial}{\partial v}\right),$$

which act on $\mathbb{C}$-valued functions in the natural way. For example, if $f(u, v) = \varphi(u, v) + i\psi(u, v)$, then

$$\frac{\partial f}{\partial z} = \frac{1}{2}\left(\frac{\partial}{\partial u} - i\frac{\partial}{\partial v}\right)(\varphi + i\psi) = \frac{1}{2}\left(\frac{\partial \varphi}{\partial u} + \frac{\partial \psi}{\partial v}\right) + \frac{i}{2}\left(\frac{\partial \psi}{\partial u} - \frac{\partial \varphi}{\partial v}\right).$$

EXERCISE 1. Prove that if $P(z) = a_n z^n + \cdots + a_0$ is a polynomial in the complex variable z, then

$$\frac{\partial P}{\partial z} = n a_n z^{n-1} + \cdots + a_1, \qquad \frac{\partial P}{\partial \overline{z}} = 0.$$

We note that the operators $\partial/\partial z$ and $\partial/\partial \overline{z}$ are a basis of the complexified tangent space ${}^{\mathbb{C}}T_P^*U$, $P \in U$ (we recall that the operators $\partial/\partial u$ and $\partial/\partial v$ can be regarded as a basis of the tangent space $T_P U$).

Consider the complex differentials $dz = du + i dv$ and $d\overline{z} = du - i dv$. We note that the complex differentials dz and $d\overline{z}$ form a basis of the adjoint (over $\mathbb{C}$) space ${}^{\mathbb{C}}T_P^*U$ to the tangent space ${}^{\mathbb{C}}T_P U$ (we recall that the differentials du and dv can be regarded as linear functionals defined on the tangent space $T_P U$:

$$du\left(a\frac{\partial}{\partial u} + b\frac{\partial}{\partial v}\right) = a, \qquad dv\left(a\frac{\partial}{\partial u} + b\frac{\partial}{\partial v}\right) = b,$$

that is, as a dual basis of the adjoint space $T_P^* U$).

PROBLEM 2. Prove that the pair $dz, d\overline{z}$ is the dual basis of the basis $\partial/\partial z$, $\partial/\partial \overline{z}$.

PROBLEM 3. Prove that the differential of a $\mathbb{C}$-valued function f is equal to

$$df = \frac{\partial f}{\partial u}du + \frac{\partial f}{\partial v}dv = \frac{\partial f}{\partial z}dz + \frac{\partial f}{\partial \overline{z}}d\overline{z}.$$

Prove that the Taylor series in complex form is

$$\begin{aligned} f(u, v) = f(z, \overline{z}) = f(z_0, \overline{z}_0) &+ \frac{\partial f(z_0, \overline{z}_0)}{\partial z}(z - z_0) + \frac{\partial f(z_0, \overline{z}_0)}{\partial \overline{z}}(\overline{z} - \overline{z}_0) \\ &+ \frac{1}{2!}\left(\frac{\partial^2 f(z_0, \overline{z}_0)}{\partial z^2}(z - z_0)^2 + 2\frac{\partial^2 f(z_0, \overline{z}_0)}{\partial z \partial \overline{z}}(z - z_0)(\overline{z} - \overline{z}_0)\right. \\ &\left. + \frac{\partial^2 f(z_0, \overline{z}_0)}{\partial \overline{z}^2}(\overline{z} - \overline{z}_0)^2\right) + \cdots, \end{aligned}$$

as if z and $\overline{z}$ are independent coordinates in the domain U.

The complex notation is often very convenient. For example, it is easy to verify that the Cauchy-Riemann conditions for a function f to be holomorphic are equivalent to $\partial f/\partial \bar{z} = 0$. Therefore, if $\partial f/\partial \bar{z} = 0$, then f is holomorphic and its Taylor series does not contain $\bar{z}$ (in this case we say that f does not depend on $\bar{z}$ and write $f(z)$). It is convenient to write the Euclidean Laplacian in terms of complex derivatives. It is easy to see that $\Delta = 4(\partial/\partial z)(\partial/\partial \bar{z})$.

We now return to minimal surfaces. Suppose that a minimal surface M in $\mathbb{R}^3$ is specified locally in conformal coordinates (u, v) and that $z = u + iv$ is the corresponding complex coordinate. Let

$$r(z, \bar{z}) = (x^1(z, \bar{z}), x^2(z, \bar{z}), x^3(z, \bar{z}))$$

be the radius vector that specifies M, where (x^1, x^2, x^3) are the standard Euclidean coordinates in $\mathbb{R}^3$. If we regard the functions $x^k(z, \bar{z})$ as complex-valued, but taking real values, we can define $\partial x^k/\partial z = \frac{1}{2}(x^k_u - ix^k_v)$. Let $\varphi = \partial r/\partial z = (x^1_z, x^2_z, x^3_z)$. Then

1) $(\varphi)^2 = \sum_{k=1}^3 (x^k_z)^2 = 0$, since $(\varphi)^2 = (|r_u|^2 - |r_v|^2) - 2i\langle r_u, r_v\rangle$, $z = u + iv$ and (u, v) are conformal coordinates. In general, if (u, v) are arbitrary coordinates on an arbitrary surface M, then $(\partial r/\partial z)^2 = 0$ is the condition for the coordinates (u, v) to be conformal, where $z = u + iv$.

2) $|\varphi|^2 = |r_u|^2 + |r_v|^2$. Therefore $|\varphi|^2 \neq 0$ is the condition for the surface M to be regular, that is, for r_u and r_v to be linearly independent. This is the regularity condition in conformal coordinates.

3) $\partial\varphi/\partial\bar{z} = \partial^2 r/\partial\bar{z}\partial z = \frac{1}{4}\Delta r = 0$. Therefore φ is a holomorphic radius vector (all the coordinates of the function $\varphi^k = \partial x^k/\partial z$ are holomorphic). We have seen (Corollary 1) that in isothermal coordinates the minimality of a surface is equivalent to the harmonicity of the radius vector that specifies this surface.

COROLLARY 4. *Let M be an arbitrary surface in $\mathbb{R}^3$ specified locally by the radius vector $r(u, v)$. Put $\varphi = \partial r/\partial z$, where $z = u + iv$. Suppose that $(\varphi)^2 = 0$. Then the fact that M is minimal is equivalent to φ being holomorphic.*

Clearly, the functions $x^k(z, \bar{z})$ can be restored by integrating the holomorphic functions φ^k. More precisely, if the domain U of the coordinate z is simply-connected, then $x^k(z, \bar{z}) = \overline{\overline{C}}_k + 2\,\mathrm{Re}\int_{z_0}^z \varphi^k\,dz$, $c_k = \text{const}$, $c_k \in \mathbb{R}$, where integration is carried out along an arbitrary piecewise-smooth path in U joining some fixed point z_0 to the point z. Since U is simply-connected and $\varphi^k(z)$ are holomorphic, the integral does not depend on the path of integration.

By Corollary 4 all minimal surfaces can be described locally (in a simply-connected domain) by means of a triple of holomorphic functions $(\varphi^1, \varphi^2, \varphi^3)$ satisfying the condition $\sum_{k=1}^3 (\varphi^k)^2 = 0$. The minimal surface is restored by integrating the functions φ^k, $x^k(z, \bar{z}) = c_k + 2\operatorname{Re}\int \varphi^k\, dz$.

By means of such a representation we can obtain a one-parameter associated family of minimal surfaces, a classical example of which is the family of isometric minimal surfaces joining the catenoid and helicoid (see §2 of Chapter 2). Let $r(z, \bar{z})$ be the radius vector that locally specifies a minimal surface M, and $\varphi = \partial r/\partial z$ the corresponding holomorphic radius vector, $(\varphi)^2 = 0$. Consider the one-parameter family of holomorphic radius vectors

$$\varphi_\theta = e^{i\theta}\varphi = (e^{i\theta}\varphi^1, e^{i\theta}\varphi^2, e^{i\theta}\varphi^3).$$

Clearly, $(\varphi_\theta)^2 = \sum_{k=1}^3 (e^{i\theta}\varphi^k) = 0$. Therefore on the surfaces M_θ: $x_\theta^k(z, \bar{z}) = c_k + 2\operatorname{Re}\int \varphi_\theta^k\, dz$ the coordinates (u, v), $z = u + iv$, are conformal, and so, since φ_θ is holomorphic, all these surfaces are minimal.

DEFINITION. The minimal surfaces $M_{\pm\pi/2}$ are said to be *conjugate* to $M_0 = M$.

We recall that the catenoid and helicoid have the following representation:

$$r_1(u, v) = \operatorname{Re}(\cosh z, -i\sinh z, z) = \operatorname{Re}\int \varphi_1 \quad \text{(catenoid)};$$

$$\begin{aligned} r_2(u, v) &= \operatorname{Im}(\cosh z, -i\sinh z, z) = -\operatorname{Re}\int i\varphi_1 \\ &= \operatorname{Re}\int e^{-i\pi/2}\varphi_1 \quad \text{(helicoid)}. \end{aligned}$$

The associated family is thus described by

$$\begin{aligned} r(u, v, \theta) &= \operatorname{Re}\int e^{-i\theta}\varphi_1 = \cos\theta \operatorname{Re}\int \varphi_1 + \sin\theta \operatorname{Im}\int \varphi_1 \\ &= \cos\theta \cdot r_1(u, v) + \sin\theta \cdot r_2(u, v) \end{aligned}$$

(compare with §2 of Chapter 2).

§3. The Gaussian mapping and the Weierstrass representation

In this section we investigate the Gaussian mapping of minimal surfaces and, using the resulting information, we construct a geometrically intuitive local representation of each minimal surface by a pair $(f(w), g(w))$ of complex-valued functions (the so-called *Weierstrass representation*). In fact, this pair of functions is one of the possible ways of writing the solution of the equation $(\varphi)^2 = 0$, where $\varphi = \partial r/\partial z$ is a holomorphic radius vector (see §2). Many geometric characteristics of minimal surfaces can be expressed in terms of the functions (f, g) of the Weierstrass representation, namely the metric, the Gaussian curvature, the Gaussian mapping, and so on. The Weierstrass representation gives the possibility of constructing an enormous number of new interesting examples of minimal surfaces, and is essentially the main research tool in the theory of two-dimensional minimal surfaces in $\mathbb{R}^3$.

1. Let M be a two-dimensional surface in $\mathbb{R}^3$. Then the Gaussian mapping n of the surface M is a mapping of M into the unit sphere S^2, $n\colon M \to S^2$, that takes each point $P \in M$ into one of the two unit normals to M at this point P. Henceforth we assume that either M is orientable, or the discussions are local; this gives us the possibility of choosing one of the two normals in a continuous way.

PROPOSITION 1. *Let M be a minimal surface in $\mathbb{R}^3$, and $n\colon M \to S^2$ the Gaussian mapping of M. Then the tangent mapping n_* in each tangent plane T_PM preserves angles between vectors and is therefore a composition of an orthogonal transformation and an expansion. If we introduce an orientation on M (locally in the case of an unorientable M) that is positive with respect to the normal n, and an orientation on S^2 that is positive with respect to the inward normal, then n_* preserves the orientation. Such mappings are said to be conformal.*

REMARK. The orientation of S^2 is usually chosen to be positive with respect to the outward normal. In this case the Gaussian mapping reverses the orientation and is said to be *anticonformal*.

PROOF. We recall that the tangent mapping n_* is defined as the mapping $n_*\colon T_PM \to T_{n(P)}S^2$, which is linear on each tangent plane T_PM to M and assigns to each tangent vector $v \in T_PM$, which is the tangent vector of a curve $\gamma(t)$ on M, $\gamma(0) = P$, $\dot\gamma(0) = v$, the tangent vector to the curve $n(\gamma(t))$ on S^2 at the point $n(P) = n(\gamma(0))$. To show that n_* preserves angles, we choose isothermal coordinates (u, v) on M in which the metric induced from $\mathbb{R}^3$ has the form $ds^2 = \lambda(u, v)(du^2 + dv^2)$. The Gaussian mapping is written as $n\colon (u, v) \mapsto n(u, v)$. Therefore under the Gaussian mapping the tangent vectors r_u and r_v go into n_u and n_v. Since r_u and r_v are perpendicular and have the same length, $|r_u|^2 = |r_v|^2 = \lambda$, it is sufficient to show that n_u and n_v are also perpendicular and have the same length (from this it follows that the tangent mapping in an orthogonal basis is given by a diagonal matrix, and so it preserves angles):

$$|n_u|^2 = \langle n_u, n_u\rangle = -\langle n_{uu}, n\rangle,$$
$$|n_v|^2 = -\langle n_{vv}, n\rangle,$$
$$\langle n_u, n_v\rangle = -\langle n_{uv}, n\rangle.$$

(All these equalities follow from the fact that $\langle n, n\rangle = 1$, so, for example, $\langle n_u, n\rangle = 0$ and $\langle n_{uu}, n\rangle + \langle n_u, n_u\rangle = 0$.) Since n_u and n_v are perpendicular to n, they can be expressed in terms of the basis r_u, r_v of the tangent space. Since

$$\langle n_u, r_u\rangle = -\langle n, r_{uu}\rangle = -b_{11}, \qquad \langle n_u, r_v\rangle = -\langle n, r_{uv}\rangle = -b_{12},$$

where

$$Q = \begin{pmatrix} b_{11} & b_{12} \\ b_{12} & b_{22} \end{pmatrix}$$

is the matrix of the second fundamental form in the basis (r_u, r_v), we have

$$n_u = -\frac{b_{11}}{\lambda} r_u - \frac{b_{12}}{\lambda} r_v.$$

Similarly,

$$n_v = -\frac{b_{12}}{\lambda} r_u - \frac{b_{22}}{\lambda} r_v.$$

Differentiating each of the equalities with respect to u and v, taking account of the fact that $\langle r_u, n\rangle = \langle r_v, n\rangle = 0$, we obtain

$$|n_u|^2 = -\langle n_{uu}, n\rangle = \frac{b_{11}^2}{\lambda} + \frac{b_{12}^2}{\lambda},$$

$$|n_v|^2 = -\langle n_{vv}, n\rangle = \frac{b_{12}^2}{\lambda} + \frac{b_{22}^2}{\lambda},$$

$$\langle n_u, n_v\rangle = -\langle n_{uv}, n\rangle = \frac{b_{11}b_{12}}{\lambda} + \frac{b_{12}b_{22}}{\lambda}.$$

Since M is a minimal surface and (u, v) are isothermal coordinates, the mean curvature $H = (b_{11} + b_{22})/2\lambda = 0$, that is, $b_{11} = -b_{22}$. Therefore, $|n_u|^2 = |n_v|^2$ and $\langle n_u, n_v\rangle = 0$. This proves that the Gaussian mapping preserves angles.

Let us consider what happens to the orientation under the Gaussian mapping. We observe that n is also the outward normal to the sphere S^2. On transition from the basis (r_u, r_v) to the basis (n_u, n_v) (so long as n_u and n_v are nonzero) the transition matrix (see above) is equal to

$$\frac{1}{\lambda}\begin{pmatrix} -b_{11} & -b_{12} \\ -b_{12} & -b_{22} \end{pmatrix},$$

and its determinant is equal to $\det Q/\lambda^2 = K$, where K is the Gaussian curvature. Since the Gaussian curvature is equal to the product of the principal curvatures, and for a minimal surface the principal curvatures have opposite signs (or they are both zero), we have $K \leq 0$. Therefore the orientations of the bases (r_u, r_v) and (n_u, n_v) are opposite. Hence the orientation of (r_u, r_v, n) coincides with the orientation of $(n_u, n_v, -n)$, where $(-n)$ is the inward normal of the sphere S^2.

REMARK. On the way we have proved that the area of the square spanned by the tangent vectors to M (in isothermal coordinates) under the Gaussian map is changed by a factor $-K$.

On the sphere S^2 we can introduce the canonical coordinates given by stereographic projection. It turns out that these coordinates are isothermal. We recall that the stereographic projection π_N of the sphere $S^2 = \{x^2 + y^2 + z^2 = 1\}$ from the North Pole $N = (0, 0, 1)$ onto the xy-plane is specified as follows. Suppose that $P \in S^2$, $P \neq N$; then $\pi_N(P) \in Oxy$ is the point of intersection of the line NP with the xy-plane. Thus, the xy-plane parametrizes the whole sphere S^2 except for the North Pole N.

Proposition 2. *The stereographic projection $\pi_N \colon S^2 \to Oxy$ of the sphere S^2 onto the xy-plane from the North Pole N preserves the angles between tangent vectors. Thus, the coordinates specified on $S^2 \backslash N$ by the stereographic projection π_N are isothermal.*

Proof. Choose an arbitrary point $P \in S^2$, $P \neq N$. If $P = S = (0, 0, -1)$ is the South Pole, everything is obvious. Suppose that $P \neq S$. In $T_P S^2$ we choose two unit tangent vectors e_1 and e_2 such that e_1 is directed along the parallel l, and e_2 along the meridian m. Since e_1, e_2 is an orthonormal basis, it is sufficient to show that $\pi_* e_1$ is orthogonal to $\pi_* e_2$ and $|\pi_* e_1| = |\pi_* e_2|$. Here π_* denotes the corresponding tangent mapping. Clearly, $\pi_* e_1$ is orthogonal to $\pi_* e_2$, since the image of a meridian is a straight line in the xy-plane passing through the origin $O = (0, 0)$, and the image of a parallel is a circle in the xy-plane with center at O. Let (ρ, φ, θ) be the standard spherical coordinates in $\mathbb{R}^3$. Then the meridian m can be parametrized by the angle θ, and the parallel l by the angle φ. Clearly, the tangent vector $\dot{m}$ has unit length, and $\dot{l}$ has length $\sin\theta$. The point $\pi_* P$ is at a distance $\operatorname{cotan}(\theta/2)$ from the origin O. Therefore along the image of the meridian $\pi(m)$ a point moves with speed $(\pi m)^{\cdot}$ equal in modulus to $|(\operatorname{cotan}(\theta/2)^{\cdot}| = 1/2\sin^2(\theta/2)$. Along the image of the parallel—a circle of radius $\operatorname{cotan}(\theta/2)$—a point moves with speed $\operatorname{cotan}(\theta/2)$. Therefore under stereographic projection the unit tangent vector along the parallel goes into a vector of length $\operatorname{cotan}(\theta/2)/\sin\theta = 1/2\sin^2(\theta/2)$ (since the tangent mapping is linear). This proves that the images of the unit vectors e_1 and e_2 have the same length.

Now a few words about the orientation of S^2. We observe that under a motion of the point P along the meridian in the direction of increasing θ, that is, from the North Pole N, the image $\pi_*(P)$ moves towards the origin O of the xy-plane. It is now clear that by means of the stereographic projection π_N the positive orientation of the xy-plane specifies on S^2 the orientation that is positive with respect to the inward normal.

Remark. Instead of the North Pole $N = (0, 0, 1)$ of the sphere S^2, we can choose the South Pole $S = (0, 0, -1)$ and consider the stereographic projection π_S onto the xy-plane from the South Pole S. On the sphere S^2 with the poles N and S removed there arise two coordinate systems. Let $\xi = x + iy$ be the complex coordinate in the xy-plane. Then it is easy to see that we can go over from one coordinate system to the other by means of the mapping $\xi \to 1/\bar{\xi}$.

If we introduce the coordinate $\xi = x + iy$ on $S^2 \backslash N$, and the coordinate $\eta = x - iy$ on $S^2 \backslash S$, then the transition function will be complex-analytic: $\eta = 1/\xi$. These are the standard coordinates on the Riemann sphere $\mathbb{C} \cup \{\infty\}$ in complex analysis.

COROLLARY. *If the sphere S^2 is regarded as the Riemann sphere $S^2 \approx \mathbb{C} \cup \{\infty\}$ by means of stereographic projection, then the Gaussian mapping n of a minimal surface M, $n\colon M \to S^2$, is represented by a meromorphic function in every isothermal coordinate system (u, v), $w = u + iv$, on M. In the coordinates on S^2 given by the stereographic projection π_N from the North Pole, the poles of n are all the points $P \in M$ at which the normal $n(P)$ has the direction of the z-axis, that is, $n(P) = N = (0, 0, 1)$. The zeros of n are those points P at which the normal $n(P)$ has the opposite direction to the z-axis, that is, $n(P) = S = (0, 0, -1)$. If we assign coordinates on S^2 by means of the stereographic projection π_S from the South Pole, then the zeros and poles of n in the previous coordinate system change places.*

2. Let (u, v) be isothermal coordinates on a minimal surface M, let $w = u + iv$ be the corresponding complex coordinate, let $r(u, v)$ be the radius vector that specifies M locally, and let $\varphi = \partial r/\partial w$ be the corresponding holomorphic radius vector, $(\varphi)^2 = 0$.

EXERCISE. Express the Gaussian mapping in terms of the components of $\varphi = (\varphi_1, \varphi_2, \varphi_3)$.

SOLUTION. We can write the unit normal n compatible with the orientation of M as $[r_u, r_v]/|[r_u, r_v]|$, where $[\ ,\ ]$ denotes the vector product.

Suppose that $[r_u, r_v] = (\Delta^1, \Delta^2, \Delta^3)$. Clearly,

$$\begin{aligned}\Delta^1 &= y_u z_v - y_v z_u \\ &= \operatorname{Im}[(y_u - iy_v)(z_u + iz_v)] = 4\operatorname{Im}(\varphi_2\overline{\varphi}_3).\end{aligned}$$

We obtain similar expressions for Δ^2 and Δ^3. Thus,

$$\begin{aligned}[r_u, r_v] &= 4\operatorname{Im}(\varphi_2\overline{\varphi}_3, \varphi_3\overline{\varphi}_1, \varphi_1\overline{\varphi}_2) \\ &= 2\begin{vmatrix} i & j & k \\ \varphi_1 & \varphi_2 & \varphi_3 \\ \overline{\varphi}_1 & \overline{\varphi}_2 & \overline{\varphi}_3 \end{vmatrix} = 2[\varphi, \overline{\varphi}].\end{aligned}$$

Moreover,

$$|[r_u, r_v]| = \lambda = |r_u|^2 = |r_v|^2 = \frac{|r_u|^2 + |r_v|^2}{2} = 2|\varphi|^2.$$

Therefore,

$$n = \frac{2\operatorname{Im}(\varphi_2\overline{\varphi}_3, \varphi_3\overline{\varphi}_1, \varphi_1\overline{\varphi}_2)}{|\varphi|^2} = \frac{[\varphi, \overline{\varphi}]}{|\varphi|^2}.$$

We now consider the stereographic projection $\pi_N\colon S^2 \to Oxy$ of the sphere S^2 onto the xy-plane and let $\xi = x + iy$ be the complex coordinate in the xy-plane. It is easy to show that if (x, y, z) are the coordinates of a point P lying on the sphere S^2, then the coordinates of its projection $\pi_N(P)$ on the xy-plane are equal to $(x/(1-z), y/(1-z))$ (see Figure 37).

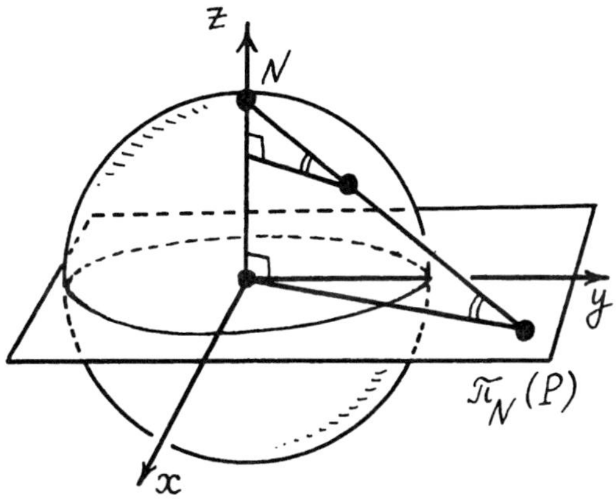

FIGURE 37

Therefore for the Gaussian mapping of the minimal surface we have

$$\xi = \frac{2\,\mathrm{Im}(\varphi_2\overline{\varphi}_3) + 2i\,\mathrm{Im}(\varphi_3\overline{\varphi}_1)}{|\varphi|^2 - 2\,\mathrm{Im}(\varphi_1\overline{\varphi}_2)}.$$

This expression can be greatly simplified.

Let us calculate the numerator of the expression for ξ:

$$\begin{aligned} &2\,\mathrm{Im}(\varphi_2\overline{\varphi}_3) + 2i\,\mathrm{Im}(\varphi_3\overline{\varphi}_1) \\ &\quad = \frac{1}{i}(\varphi_2\overline{\varphi}_3 - \overline{\varphi}_2\varphi_3 + i\varphi_3\overline{\varphi}_1 - i\overline{\varphi}_3\varphi_1) \\ &\quad = \varphi_3(\overline{\varphi}_1 + i\overline{\varphi}_2) - \overline{\varphi}_3(\varphi_1 + i\varphi_2). \end{aligned}$$

Using the fact that

$$\varphi_1^2 + \varphi_2^2 + \varphi_3^2 = (\varphi_1 - i\varphi_2)(\varphi_1 + i\varphi_2) + \varphi_3^2 = 0,$$

that is, $\varphi_1 + i\varphi_2 = -\varphi_3^2/(\varphi_1 - i\varphi_2)$ we have

$$\begin{aligned} &\varphi_3(\overline{\varphi}_1 + i\overline{\varphi}_2) + \overline{\varphi}_3 \cdot \frac{\varphi_3^2}{\varphi_1 - i\varphi_2} \\ &\quad = \frac{\varphi_3[(\overline{\varphi}_1 + i\overline{\varphi}_2)(\varphi_1 - i\varphi_2) + |\varphi_3|^2]}{\varphi_1 - i\varphi_2} \\ &\quad = \frac{\varphi_3}{\varphi_1 - i\varphi_2}[|\varphi|^2 - 2\,\mathrm{Im}(\varphi_1\overline{\varphi}_2)]. \end{aligned}$$

Thus $\xi = \varphi_3/(\varphi_1 - i\varphi_2)$ is the meromorphic function corresponding to the Gaussian mapping.

Let $g = \varphi_3/(\varphi_1 - i\varphi_2)$. We put $f = \varphi_1 - i\varphi_2$. The function f is obviously holomorphic. Then we can express the three components of φ in terms of f and g:

$$\varphi_1 = \frac{1}{2}f(1 - g^2), \quad \varphi_2 = \frac{i}{2}f(1 + g^2), \quad \varphi_3 = fg,$$

and since $\varphi_1, \varphi_2, \varphi_3$ are holomorphic functions, the product fg^2 must be holomorphic.

DEFINITION. A pair of complex-valued functions (f, g), defined on a simply-connected domain U of the complex plane $\mathbb{C}$, $U \subset \mathbb{C}$, such that f is holomorphic, g is meromorphic, and fg^2 is holomorphic, is called a *Weierstrass representation.*

Thus, we have proved the following theorem.

THEOREM. *Let (f, g) be a Weierstrass representation in a (simply-connected) domain U of the complex plane $\mathbb{C}$. If we form three holomorphic functions $\varphi_1 = \frac{1}{2}f(1-g^2)$, $\varphi_2 = \frac{i}{2}f(1+g^2)$, and $\varphi_3 = fg$, then the mapping of U into $\mathbb{R}^3$ given by*

$$x^k(w, \overline{w}) = c_k + 2\,\mathrm{Re}\int_{w_0}^{w} \varphi_k\, dw$$

where $|\varphi|^2 \neq 0$ determines a regular immersed (that is, with possible self-intersections) minimal surface, and from the fact that $(\varphi)^2 = \sum_{k=1}^{3} \varphi_k^2 = 0$, it follows that the coordinates (u, v), $w = u + iv$, are isothermal (if $r(u, v) = (x^1(u, v), x^2(u, v), x^3(u, v))$, then $\varphi = \partial r/\partial w$). Here $\{x^k\}_{k=1}^3$ are the standard Euclidean coordinates in $\mathbb{R}^3$., and w is the complex coordinate in $U \subset \mathbb{C}$.

Conversely, if (u, v) are conformal coordinates on M in a simply-connected domain $U \subset \mathbb{C}$, $w = u + iv$ is the corresponding complex coordinate, $r(u, v)$ is the radius vector that specifies M locally, and $\varphi = \partial r/\partial w$ is the corresponding holomorphic radius vector with components $(\varphi_1, \varphi_2, \varphi_3)$, then by putting $f = \varphi_1 - i\varphi_2$ and $g = \varphi_3/(\varphi_1 - i\varphi_2)$ we obtain a Weierstrass representation of the minimal surface M in the domain $U \subset \mathbb{C}$. The function g is a Gaussian mapping of the minimal surface M if on S^2 the coordinates are given by the stereographic projection π_N from the North Pole.

It is useful to express the main geometric characteristics of a minimal surface in terms of the functions of the Weierstrass representation.

1) $\lambda = 2|\varphi|^2 = |f|^2(1 + |g|^2)^2$ where λ is the conformal factor of the induced metric on M in the coordinates given by the Weierstrass representation. Therefore a sufficient condition for such a minimal surface to be regular is $f \neq 0$. Since g is meromorphic, it is possible that $f(w_0) = 0$ but $fg^2(w_0) \neq 0$.

If we allow the existence of singular points on minimal surfaces, then such surfaces will be called *generalized minimal surfaces.* Singular points of generalized minimal surfaces have a special form (this follows from the Weierstrass representation), and so they have acquired a special name—they are called *branch points.* The condition that w_0 is a branch point is equivalent to $f(w_0) = 0$ and $fg^2(w_0) = 0$ simultaneously.

2) *The Gaussian curvature* K. As we mentioned after the proof of Proposition 1, the area of the square spanned by (r_u, r_v) under the Gaussian mapping $(r_u, r_v) \to (n_u, n_v)$ is increased by a factor of $-K$. Since the area of the square (r_u, r_v) is equal to λ, the area of the square (n_u, n_v) is equal to $-K\lambda$.

Next, suppose that the coordinate $w = u + iv$ of the point $P \in M$ is not a pole of the meromorphic function g from the Weierstrass representation (f, g) (this means that $n(P) \neq N$). Then under the stereographic projection π_N the square (n_u, n_v) tangent to S^2 at the point $n(P)$ goes into the square $(\pi_{N_*} n_u, \pi_{N_*} n_v)$ on the xy-plane. Clearly, the area of the square $(\pi_{N_*} n_u, \pi_{N_*} n_v)$ is equal to $|\dot{g}(w)|^2$, where $\dot{g} = dg/dw$.

We already know that the tangent mapping π_{N_*} preserves angles, and so it expands each vector by the same factor. Therefore, to determine the coefficient of expansion it is sufficient to calculate it on any vector. Let us choose the unit tangent vector to the meridian. If (φ, θ) are the coordinates on S^2 induced by the spherical coordinate system (r, φ, θ) in $\mathbb{R}^3$, then, as we showed in the proof of Proposition 1, this unit vector is expanded $1/2\sin^2(\theta/2)$ times, and the distance from $\pi_N(n(P))$ to 0 is equal to $\operatorname{cotan}(\theta/2)$. If $\xi = x + iy$ and $|\xi| = \operatorname{cotan}(\theta/2)$, then the coefficient of expansion is equal to $1/2\sin^2(\theta/2) = (1+|\xi|^2)/2$. Putting $\xi = g(w)$, we see that the area of the square (n_u, n_v) is equal to $-K\lambda = |\dot{g}|^2/((1+|g|^2)/2)^2$. Consequently, $K = -4|\dot{g}|^2/|f|^2(1+|g|^2)^4$.

Let us recall that a point of a regular surface is said to be an *umbilic* if the two principal curvatures are equal there. On a minimal surface this means that the two principal curvatures are both zero, or equivalently that the Gaussian curvature K is zero. From the formula just given for the Gaussian curvature it follows that umbilics of a minimal surface M are just the zeros of the derivative of the function g in the Weierstrass representation (f, g) for M.

Later the expression for the Gaussian curvature will be useful to us in the calculation of the indices of minimal surfaces.

3. Let us give examples of the Weierstrass representation for the classical minimal surfaces.

The simplest representation $(1, w)$ on the whole plane, $U = \mathbb{C}$, gives the so-called Enneper surface. The explicit equations of the Enneper surface are

$$x = \rho\cos\varphi - \frac{\rho^3}{3}\cos 3\varphi,$$

$$y = -\rho\sin\varphi - \frac{\rho^3}{3}\sin 3\varphi,$$

$$z = \rho^2\cos 2\varphi,$$

where (ρ, φ) are polar coordinates in the w-plane, $\rho = |w|$, $\varphi = \arg w$.

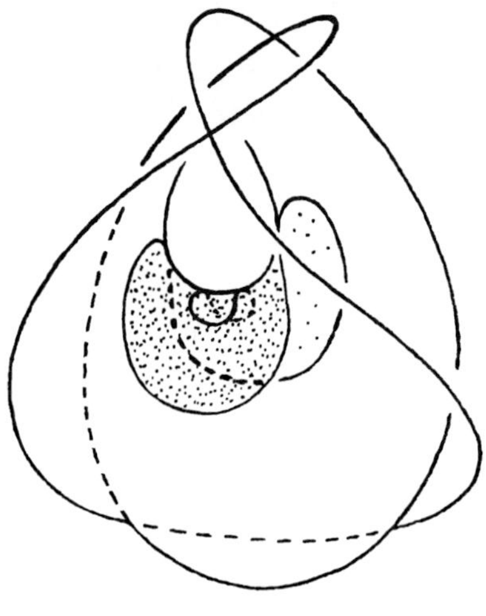

FIGURE 38

This classical minimal surface has a saddle shape in the neighborhood of the origin. As we increase the domain $U_r = \{w \mid |w| < r\}$ the upper and lower ends of the contour Γ_r—the image of ∂U_r—begin to come together and at a certain instant a self-intersection of the surface occurs (see Figure 38).

The next classical example is the catenoid $r = \cosh z$. Its Weierstrass representation is $(-1/2w^2, w)$ on $\mathbb{C}\backslash\{0\}$. The domain $\mathbb{C}\backslash\{0\}$ is not simply-connected. However, we can still define the Weierstrass representation in this case. We needed the domain to be simply-connected in order that the integral $\int_{w_0}^{w} \varphi_k\, dw$ should not depend on the path joining w_0 and w. For a multiply-connected domain we define the so-called periods, that is, integrals over closed piecewise-smooth curves that are not contractible to a point by any continuous deformation. The real (imaginary) part of a period is called the real (respectively, imaginary) period.

In order for the Weierstrass representation in a multiply-connected domain to correctly define a minimal surface it is necessary and sufficient that all the real periods should be equal to zero, that is, that the integrals $\int \varphi_k\, dw$ along any closed contour should be purely imaginary numbers.

It is easy to verify that this property is satisfied in the case of the Weierstrass representation for the catenoid. If (ρ, φ) are polar coordinates in the w-plane, then the catenoid is given by

$$x = \frac{1}{2}\left(\frac{1}{\rho} + \rho\right)\cos\varphi, \quad y = \frac{1}{2}\left(\frac{1}{\rho} + \rho\right)\sin\varphi, \quad z = -\ln\rho.$$

Thus, the coordinates (ρ, φ) are connected with the coordinates (z, φ) described in §1 of Chapter 2 by the formulas $\rho = e^{-z}$, $\varphi = \varphi$. As $\rho \to 0$ the coordinate $z \to +\infty$. The parallels $z = \text{const}$ correspond in the w-plane to concentric circles $\rho = |w| = \text{const}$, which contract to the point $\rho = 0$ as $z \to +\infty$.

If we slightly vary the Weierstrass representation for the catenoid by putting $(-i/2w^2, w)$, then on $\mathbb{C}\backslash\{0\}$ the resulting representation will have real periods. To get rid of them, we cut out from plane $\mathbb{C}$ the nonpositive real axis $\mathbb{R}_- = \{u \leq 0, v = 0\}$. The domain $\mathbb{C}\backslash\mathbb{R}_-$ is simply-connected. It is easy to see that we have obtained one coil of the helicoid

$$x = -\frac{1}{2}\left(\rho - \frac{1}{\rho}\right)\sin\varphi = \sinh a \sin b,$$

$$y = \frac{1}{2}\left(\rho - \frac{1}{\rho}\right)\cos\varphi = -\sinh a \cos b,$$

$$z = \varphi = b,$$

where $|\varphi| < \pi$, $\rho > 0$, and $\rho = e^{-a}$, $\varphi = b$.

If we now glue together countably many copies of $\mathbb{C}\backslash\mathbb{R}$ in the natural way, that is, the upper side of the kth copy to the lower side of the $(k+1)$st, $k \in \mathbb{Z}$, then we obtain a complete helicoid. This is equivalent to allowing the angle φ to vary from $-\infty$ to $+\infty$. We naturally wish to consider the change of coordinates specified by the exponential e^w.

In fact, the complete helicoid can be specified by another Weierstrass representation, namely, $(\frac{-i}{2}e^{-w}, e^w)$ on $\mathbb{C}$:

$$x = -\sinh u \sin v, \quad y = \sinh u \cos v, \quad z = v$$

where $w = u + iv$.

It is easy to verify that the Weierstrass representation (e^{-w}, e^w) on $\mathbb{C}$ specifies an infinite-sheeted covering of the catenoid. In the (u, v)-coordinates, where $u + iv = w$, the winding (covering) is given by

$$x = -\cosh u \cos v, \quad y = -\cosh u \sin v, \quad z = u.$$

REMARK. If (f, g) is a Weierstrass representation, then the associated family (see §2 of Chapter 2) is defined as the family of minimal surfaces with Weierstrass representation $(e^{i\theta}f, g)$. The conjugate surfaces have the representation $(\pm if, g)$ (compare with the Weierstrass representation for the catenoid and helicoid).

ASSERTION. *The associated family for a minimal surface M consists of locally isometric minimal surfaces (as a rule, pairwise noncongruent).*

PROOF. Let $(e^{i\theta}f(w), g(w))$ be the Weierstrass representation for the surface M_θ of the associated family, where $w = u + iv$. Then the metric on M_θ induced from $\mathbb{R}^3$ has the form $ds^2 = \lambda(\theta, u, v)(du^2 + dv^2)$ and $\lambda(\theta, u, v) = |e^{i\theta}f|^2(1+|g|^2)^2 = |f|^2(1+|g|^2)^2$ that is, it does not depend on θ.

Let us give without proof the Weierstrass representations for the other classical minimal surfaces.

The Scherk surface: $(1/(1-w^4), w)$ on $U = \{|w| < 1\}$ (the graph). If we consider this representation on $\mathbb{C}\backslash\{w^4 = 1\}$, then the integrals $\int \varphi_k \, dw$ will have real periods. By analogy with what we did in the case of the helicoid, we can obtain a representation for the complete Scherk surface.

The Schwarz-Riemann surface: $(1/\sqrt{1 - 14w^4 + w^8}, w)$ on a suitable domain $U \subset \mathbb{C}$.

The Richmond surface: $(w^2, 1/w^2)$ on $\mathbb{C}\backslash\{0\}$.

§4. The global Weierstrass representation

To understand the rest of the text we need to have an idea of smooth manifolds and the basic objects connected with them: smooth functions, curves, tangent and cotangent vectors, vector fields, and differential forms (and more generally tensors). We should also understand how these objects are transformed under smooth mappings of one manifold into another. We refer those readers who are not familiar with these ideas to [1], [7], and [15].

In the previous section we showed that any minimal surface M can be specified locally by a pair of $\mathbb{C}$-valued functions (f, g), called a local Weierstrass representation. In this section we state how it is possible to "glue together" local representations corresponding to different coordinate domains of the surface M into a single (global) Weierstrass representation, which will then describe the whole minimal surface M. The idea of regarding a surface as a set (topological space) glued together from coordinate domains leads to the notion of a manifold.

Henceforth a surface will be understood as an immersion $\psi: M \to \mathbb{R}^3$ of a connected two-dimensional manifold M into $\mathbb{R}^3$. We recall that a smooth mapping $\psi: M \to N$ of a smooth m-dimensional manifold M into a smooth n-dimensional manifold N $(m < n)$ is called an *immersion* if the tangent mapping ψ_* at each point $P \in M$ is an embedding of the tangent space $T_P M$ into the tangent space $T_{f(P)} N$, that is, $\psi_*: T_P M \hookrightarrow T_{f(P)} N$. An immersion $\psi: M \to N$ is called an *embedding* if ψ establishes a homeomorphism between the manifold M and its image $\psi(M) \subset N$. Locally any immersion is an embedding (one-to-one with its image), so when the arguments have a local character we shall identify the manifold M and its image in $\mathbb{R}^3$. In this case we shall not distinguish between the tangent (cotangent) space $T_P M$ $(T_P^* M)$ and its image in $\mathbb{R}^3$, keeping the same notation for both objects. More generally, for convenience of exposition we shall talk about "a surface M", bearing in mind that we are given an immersion $\psi: M \to \mathbb{R}^3$ of the manifold M into $\mathbb{R}^3$.

All the classical minimal surfaces considered above are naturally still surfaces in the new understanding. Thus, the Enneper surface is an immersion of the plane $\mathbb{R}^2$ into $\mathbb{R}^3$, and the catenoid is an embedding of the plane with a point deleted $\mathbb{R}^2\backslash\{(0, 0)\}$ into $\mathbb{R}^3$. An important example of a manifold and surface is the standard sphere $S^2 \subset \mathbb{R}^3$. The stereographic projections φ_N and φ_S from the North and South Poles specify the coordinates on $S^2\backslash N$ and $S^2\backslash S$ respectively.

We now define more accurately what such a minimal surface is. For this we need a definition of the metric and the second fundamental form.

DEFINITION OF THE METRIC. A *metric* on a manifold M is a scalar product, defined on each tangent space T_PM and depending smoothly on the point P. A manifold on which a metric is specified is called a *Riemann manifold*.

REMARK. Generally speaking, all objects that can be defined on a vector space carry over to a manifold. For this we need to define this object on each tangent (cotangent) space and require that it depends smoothly on a point of the manifold (so that in any coordinate system all the functions that take part in its definition should be smooth). We defined the metric in this way above. Similarly we can define any bilinear form on M, tensor fields, and so on.

If $\psi\colon M \to \mathbb{R}^3$ is an immersion, we can define the metric ds^2 on M induced by this immersion: we define the scalar product $ds^2(\xi, \eta)$ of the tangent vectors ξ and η lying in T_PM as the scalar product in $\mathbb{R}^3$ of their images $\psi_*(\xi)$ and $\psi_*(\eta)$ under the action of the tangent mapping ψ_*.

We now define the second fundamental form of a surface M.

DEFINITION. The *second fundamental form* of a surface M is the vector-valued bilinear form (with values in $\mathbb{R}^3$) defined as follows. Let $v \in T_PM$ be the velocity vector of a curve $\gamma(t)$ lying on the surface M, where $\gamma(0) = P$; then the quadratic form from which we can uniquely restore the bilinear form Q is defined as $Q(v) = (d^2\gamma(t)/dt^2|_{t=0})^N$ where $(\ )^N$ denotes orthogonal projection onto a line perpendicular to T_PM (see the agreement above). In other words, the quadratic form $Q(v)$ is equal to the normal component of the acceleration vector of the curve $\gamma(t)$.

For an orientable surface (the manifold M is orientable) we can define an $\mathbb{R}$-valued second fundamental form Q. For this we choose (since the surface M is orientable) the family of unit normals $n(P)$ to M that depend smoothly on a point P. This means that there is a smooth mapping $n\colon M \to S^2$ that assigns to each point P the unit normal $n(P)$ to M. The mapping n is called the *Gaussian mapping*. We put $Q(v) = \langle d^2\gamma/dt^2|_{t=0}, n\rangle$ where $\langle\ ,\ \rangle$ is the standard scalar product in $\mathbb{R}^3$.

Let $G = (g_{ij})$ be the matrix of the metric, and $Q = (b_{ij})$ the matrix of an arbitrary bilinear form (in some coordinate chart). We form the local function $H = \operatorname{tr} G^{-1}Q = \sum g^{ij}b_{ij}$ where $(g^{ij}) = G^{-1}$ is the inverse of the matrix of the metric. It is easy to verify that the local functions H can be glued together into a single function on M. The resulting function (which we also denote by H) is called the *trace* of the bilinear form Q.

DEFINITION. The function H that is the trace of the second fundamental form Q is called the *mean curvature* (if Q is defined as a vector-valued form, then all the b_{ij} and the mean curvature are smooth families of vectors orthogonal to the surface M). The surface M is called minimal if its mean curvature H is identically zero: $H \equiv 0$.

As we mentioned in §1, on an orientable surface we can introduce a complex structure generated by systems of isothermal coordinates. A two-dimensional manifold with a complex structure is called a *Riemann surface* (not to be confused with a Riemannian manifold).

If M is a Riemann surface, then at each point $P \in M$ we can define the complex tangent space ${}^{\mathbb{C}}T_PM$ and the complex cotangent space ${}^{\mathbb{C}}T_P^*M$. The tangent space ${}^{\mathbb{C}}T_PM$ is the complexification of the real tangent space T_PM of the Riemann surface M, regarded as a real two-dimensional manifold. If (u, v) are local coordinates on M, and $\partial/\partial u$, $\partial/\partial v$ is a basis in T_PM, then the elements of ${}^{\mathbb{C}}T_PM$ are linear combinations $\alpha\partial/\partial u+\beta\partial/\partial v$ with complex coefficients, $\alpha, \beta \in \mathbb{C}$. The vectors of ${}^{\mathbb{C}}T_PM$ can be regarded as differentiations in "complex directions" of $\mathbb{C}$-valued functions defined on M (or in a neighborhood of a point $P \in M$). The cotangent space is defined as the space of all possible differentials at a point $P \in M$ of $\mathbb{C}$-valued functions on M. Each such differential determines in a natural way an $\mathbb{R}$-linear mapping of the real tangent space T_PM into complex numbers, which can be extended to a $\mathbb{C}$-linear mapping of ${}^{\mathbb{C}}T_PM$ into $\mathbb{C}$, that is, to a $\mathbb{C}$-linear functional. Therefore ${}^{\mathbb{C}}T_P^*M$ is conjugate to ${}^{\mathbb{C}}T_PM$. As before, the bases (over $\mathbb{C}$) $\partial/\partial u$, $\partial/\partial v$ for ${}^{\mathbb{C}}T_PM$ and du, dv for ${}^{\mathbb{C}}T_P^*M$ are dual. Other dual bases are

$$\frac{\partial}{\partial z} = \frac{1}{2}\left(\frac{\partial}{\partial u} - i\frac{\partial}{\partial v}\right), \qquad \frac{\partial}{\partial \bar{z}} = \frac{1}{2}\left(\frac{\partial}{\partial u} + i\frac{\partial}{\partial v}\right)$$

for ${}^{\mathbb{C}}T_PM$ and

$$dz = du + i dv, \qquad d\bar{z} = du - i dv$$

for ${}^{\mathbb{C}}T_P^*M$.

A complex-valued 1-form ω is a specification at each point $P \in M$ of a linear (over $\mathbb{C}$) mapping ω_P: ${}^{\mathbb{C}}T_PM \to \mathbb{C}$ that depends smoothly on the point. In local coordinates (u, v), $z = u + iv$, the form ω can be written as $\omega = \omega_1 dz + \omega_2 d\bar{z}$, where the $\omega_k(u, v)$ are smooth $\mathbb{C}$-valued functions of the coordinates (u, v).

A 1-form ω on a Riemann surface is said to be *holomorphic* if in every coordinate system $z = u + iv$ the form ω can be written as $\omega = \omega_1 dz$, where $\omega_1(u, v)$ is a holomorphic function, that is, $\partial\omega_1/\partial\bar{z} = 0$, $\omega_1(u, v) = \omega_1(z)$.

To construct a global Weierstrass representation, it remains to define the integral of a 1-form on M along a smooth curve in M. Let γ: $[a, b] \to M$ be a smooth curve, and ω is 1-form. Then $\gamma^*(t)$ is a family of tangent vectors, the "vector field along γ", and we can define the function $\omega(\gamma^*(t))$ on $[a, b]$. The integral of this function along $[a, b]$ is called the *integral of the* 1*-form* ω *along the curve* γ :

$$\int_\gamma \omega \stackrel{\text{def}}{=} \int_a^b \omega(\gamma^*(t))\,dt.$$

Now let $\psi\colon M \to \mathbb{R}^3$ be an immersed minimal surface, and suppose that M is orientable. Let us regard M as a Riemann surface with a complex structure induced by isothermal coordinates. If $(U, \eta\colon U \to V)$ is a coordinate chart with complex coordinate $z = u + iv$, then as in §2 we can define the holomorphic radius vector $\varphi = \partial\psi/\partial z$, where $\psi(u, v) = \psi \circ \eta^{-1}(u, v)$. We have defined a local Weierstrass representation of a pair (f, g) of $\mathbb{C}$-valued functions on $U \subset \mathbb{C}$: if $\varphi = (\varphi_1, \varphi_2, \varphi_3)$, then $f = \varphi_1 - i\varphi_2$ and $g = \varphi_3/(\varphi_1 - i\varphi_2)$.

We now form in this domain a local holomorphic 1-form $\omega = f dz$. If we now form in another coordinate system z' a local 1-form $\omega' = f' dz'$, where

$$f' = \varphi_1' - i\varphi_2' \quad \varphi' = \frac{\partial\psi}{\partial z'} = (\varphi_1', \varphi_2', \varphi_3'),$$

then in fact ω' coincides with ω in the intersection of the domains of definition of the coordinate charts. In fact, since

$$\varphi_k' = \frac{\partial \psi_k}{\partial z'} = \frac{\partial \psi_k}{\partial z}\frac{\partial z}{\partial z'},$$

we have

$$f'(z') = f(z(z'))\frac{\partial z}{\partial z'} \quad \text{and} \quad \omega' = f\frac{\partial z}{\partial z'}dz' = f dz,$$

since under a holomorphic change of coordinates $z \to z'$ the basis covector $dz \in {}^{\mathbb{C}}T_P^*M$ goes into the covector $(\partial z/\partial z')\,dz'$ (verify this).

Thus, the local forms $\omega = f dz$ are glued together into a single holomorphic 1-form on the Riemann surface M. Moreover, the locally defined functions $g = \varphi_3/(\varphi_1 - i\varphi_2)$ are also glued together into a single meromorphic function on M: if $g' = \varphi_3'/(\varphi_1' - \varphi_2')$ then in the intersection we have

$$g' = \frac{\varphi_3 \partial z'/\partial z}{\varphi_1 \partial z'/\partial z - i\varphi_2 \partial z'/\partial z} = \frac{\varphi_3}{\varphi_1 - i\varphi_2} = g.$$

This enables us to define three holomorphic 1-forms $\frac{1}{2}(1 - g^2)\omega$, $\frac{i}{2}(1 + g^2)\omega$, and $g\omega$. Locally these three forms can be written as $\varphi_1\,dz$, $\varphi_2\,dz$, and $\varphi_3\,dz$ respectively, so we denote these forms by φ^k, $k = 1, 2, 3$. If γ is an arbitrary piecewise-smooth curve joining a fixed point $P \in M$ and a point $Q \in M$, then $\psi^k(Q) = \psi^k(P) + 2\operatorname{Re}\int_\gamma \varphi^k$. It is easy to obtain this result for a curve γ lying entirely in one coordinate chart, since

$$2\operatorname{Re}\int_\gamma \varphi^k = \int_\gamma \frac{\partial\psi^k}{\partial u}du + \frac{\partial\psi^k}{\partial v}dv = \int_\gamma d\psi^k = \psi^k(Q) - \psi^k(P).$$

For an arbitrary curve γ this is obtained if we split γ into pieces, each of which lies entirely in some chart.

From what we have said it follows, in particular, that the forms φ^k do not have real periods, that is, the real part of the integral $\int \varphi^k$ along any closed piecewise-smooth curve is zero.

DEFINITION. A pair (ω, g) consisting of a holomorphic 1-form ω and a meromorphic function g on a Riemann surface M and such that the forms $\varphi^1 = \frac{1}{2}(1-g^2)\omega$, $\varphi^2 = \frac{i}{2}(1+g^2)\omega$ and $\varphi^3 = g\omega$ are holomorphic 1-forms not having real periods is called a *global Weierstrass representation.*

If we are given a global Weierstrass representation, then the mapping $\psi\colon M \to \mathbb{R}^3$ defined as $\psi^k(Q) = c_k + 2\operatorname{Re}\int_\gamma \varphi^k$, where γ is an arbitrary piecewise-smooth curve on M joining a fixed point $P \in M$ and a point $Q \in M$, and $c = (c_1, c_2, c_3) = \psi(P)$, specifies an immersed (except for a certain number of isolated points) minimal surface. Points at which the regularity of the minimal surface breaks down are called *branch points* (see §3). A point $Q \in M$ is a branch point if and only if the forms ω and $g^2\omega$ simultaneously vanish there. A minimal surface that admits branch points is called a *generalized minimal surface.*

As before, the function g specifies a Gaussian mapping of the minimal surface if we consider coordinates on the sphere S^2 obtained by means of stereographic projection from the North Pole. The poles of g are the points that are mapped into the North Pole by the Gaussian mapping, and the zeros of g are the points mapped into the South Pole.

We have thus proved the following theorem.

THEOREM. *All generalized orientable minimal surfaces in $\mathbb{R}^3$ are described by a global Weierstrass representation, that is, by all possible pairs (ω, g) consisting of a holomorphic 1-form ω and a meromorphic function g defined on a Riemann surface M and such that the forms $\varphi^1 = \frac{1}{2}(1-g^2)\omega$, $\varphi^2 = \frac{i}{2}(1+g^2)\omega$, and $\varphi^3 = g\omega$ are holomorphic and do not have real periods on M.*

The pair (ω, g) specifies a branched minimal immersion $\psi\colon M \to \mathbb{R}^3$ (an immersion having isolated branch points) as follows:

$$\psi^k = c_k + 2\operatorname{Re}\int \varphi^k, \quad \text{where } c_k = \text{const} \in \mathbb{R}$$

and the integration is carried out along an arbitrary piecewise-smooth path joining a fixed point $P \in M$ and a variable point $Q \in M$, $\psi(P) = (c_1, c_2, c_3)$.

We have already said that a minimal surface $\psi\colon M \to \mathbb{R}^3$ can also be defined in the unorientable case, that is, if M is an unorientable manifold. In this situation also it is possible to construct a global Weierstrass representation. For this we need to construct a "doubling" $\widetilde{M}$ of the manifold M, that is, an orientable two-dimensional manifold $\widetilde{M}$ together with a smooth mapping (projection) $\pi\colon \widetilde{M} \to M$, such that at each point $P \in M$ there is a sufficiently small neighborhood U whose inverse image $\pi^{-1}(U)$ consists of exactly two connected disjoint open sets V_1 and V_2 such that the restriction of π to each of them is a diffeomorphism:

$$\pi\colon V_k \overset{\text{diff}}{\approx} U, \qquad k = 1, 2.$$

In this case the immersion $\psi\colon M \to \mathbb{R}^3$ can be extended to an immersion $\widetilde{\psi}\colon \widetilde{M} \to \mathbb{R}^3$, where $\widetilde{\psi} = \psi \circ \pi$. It is convenient to represent this by the commutative diagram

$$\begin{array}{ccc} \widetilde{M} & \overset{\widetilde{\psi}}{\longrightarrow} & \mathbb{R}^3 \\ \pi\downarrow & \nearrow\psi & \\ M & & \end{array}.$$

The projection $\pi\colon \widetilde{M} \to M$ is called a *two-sheeted covering*.

Such a doubling can always be constructed. For this it is sufficient to choose a covering of M by coordinate charts (U_k, η_k) such that each U_k, and also each nonempty intersection $U_k \cap U_l$, is a connected open set (domain) (try to construct such a covering). We then take twice as big a domain $U_k^{\cdot} = \{U_k^+, U_k^-\}$, $U_k^{\pm} = U_k$, and choose coordinates (u_k, v_k) in U_k^+ and coordinates $(u_k, -v_k)$ in U_k^-, specified by means of the homeomorphisms $\eta_k^+ = \eta_k$ and $\eta_k^- = \tau \circ \eta_k$ respectively, where $\tau\colon \mathbb{R}^2 \approx \mathbb{C} \to \mathbb{R}^2 \approx \mathbb{C}$ is a conjugation. We can then glue together $U_k^{\cdot}$ and $U_l^{\cdot}$ so long as U_k intersects U_l and the transition functions $\eta_k^{\cdot}(\eta_l^{\cdot})^{-1}$ preserve the orientation (we glue along the intersection). We thus obtain an orientable manifold $\widetilde{M}$, called the doubling of M. The natural projection $\pi\colon \widetilde{M} \to M$ has the properties listed above. Moreover, on $\widetilde{M}$ there is a diffeomorphism $\widetilde{I}\colon \widetilde{M} \to \widetilde{M}$ that is an involution, that is, $\widetilde{I}^2 = \mathrm{id}$, where id is the identity transformation. The involution $\widetilde{I}$ interchanges the points of each pair—the inverse image of a point under the projection π: if $P \in M$ and $\pi^{-1}(P) = \{P_1, P_2\}$, then $\widetilde{I}(P_1) = P_2$ and $\widetilde{I}(P_2) = P_1$.

If we introduce a complex structure on $\widetilde{M}$, then the involution $\widetilde{I}$ is an antiholomorphic mapping of $\widetilde{M}$ into itself without fixed points: in each chart $\widetilde{I}$ is the composition of a holomorphic mapping and a conjugation.

Let (U, η) be a coordinate chart on $\widetilde{M}$, where U is a domain that projects one-to-one into M. If $z = u + iv$ is a coordinate in U, then in $\widetilde{I}(U)$ we can choose a coordinate $z' = v + iu$ (compatible with the complex structure).

If $\psi\colon M \to \mathbb{R}^3$ is an immersed minimal surface, then $\widetilde{\psi}\colon \widetilde{M} \to \mathbb{R}^3$ is also an immersed minimal surface, a two-sheeted covering of $\psi\colon M \to \mathbb{R}^3$. If $\varphi^1, \varphi^2, \varphi^3$ are holomorphic 1-forms specifying a global Weierstrass representation on $\widetilde{M}$ for a minimal surface $\widetilde{\psi}\colon \widetilde{M} \to \mathbb{R}^3$, then it is easy to show that $\widetilde{I}^*(\varphi^k) = \overline{\varphi}^k$, where $\widetilde{I}^*$ is the mapping of 1-forms induced by the involution $\widetilde{I}:(\widetilde{I}^*\varphi^k)(\xi) = \varphi^k(\widetilde{I}_*(\xi))$, $\xi \in^{\mathbb{C}} T_P\widetilde{M}$ is a tangent vector, and $\widetilde{I}^*$ is the tangent mapping.

Moreover, if (ω, g) is the Weierstrass representation for ψ: $\widetilde{M} \to \mathbb{R}^3$, then $g(\widetilde{I}(l)) = -1/\overline{g}(l)$ and $\widetilde{I}^*(\omega) = -\overline{g^2\omega}$. The first formula follows from the fact that g is the composition of the Gaussian mapping and a stereographic projection, and under the immersion $\widetilde{\psi}$ the points P and $\widetilde{I}(P)$ go into one point, at which the normals $n(P)$ and $n(\widetilde{I}(P))$ that orient this surface are in opposite directions: g and $-1/\overline{g}$ are the stereographic coordinates of n and $-n$ (verify this). The second formula is obtained by direct calculation (verify this, using, if necessary, the formulas

$$\frac{\varphi_1 + i\varphi_2}{\varphi_1 - i\varphi_2} = -\frac{\varphi_3^2}{(\varphi_1 - i\varphi_2)^2} = -g^2, \qquad f = \varphi_1 - i\varphi_2$$

where $\varphi = (\varphi_1, \varphi_2, \varphi_3) = \partial\psi/\partial z)$.

DEFINITION. In the case of a nonorientable two-dimensional manifold M the Weierstrass representation is (ω, g), defined on the oriented doubling $\widetilde{M}$ of M and satisfying the following additional conditions: if $\widetilde{I}$: $\widetilde{M} \to \widetilde{M}$ is an involution on $\widetilde{M}$ that interchanges the points of each pair consisting of the inverse image of a point under the canonical projection π: $\widetilde{M} \to M$, then $g(\widetilde{I}(p)) = -1/\overline{g}(p)$ and $\widetilde{I}^2(\omega) = -\overline{g^*\omega}$.

It turns out that the additional conditions on g and ω described in the definition of the Weierstrass representation in the nonorientable case completely characterize the Weierstrass representations of orientable minimal surfaces which in fact specify two-sheeted coverings of nonorientable minimal surfaces.

THEOREM. *If (ω, g) is a Weierstrass representation on a Riemann surface $\widetilde{M}$ that specifies an immersed orientable connected minimal surface $\widetilde{\psi}$: $\widetilde{M} \to \mathbb{R}^3$, and if there is an antiholomorphic involution $\widetilde{I}$: $\widetilde{M} \to \widetilde{M}$ without fixed points such that $g(\widetilde{I}(p)) = -1/\overline{g}(p)$ and $\widetilde{I}^*(\omega) = -\overline{g^2\omega}$, then $\widetilde{M}$ is a doubling of an nonorientable manifold M, π: $\widetilde{M} \to M$ is the corresponding projection, and $\psi = \widetilde{\psi} \circ \pi^{-1}$: $M \to \mathbb{R}^3$ correctly defines an immersion that specifies a connected nonorientable minimal surface.*

EXAMPLE (immersion of an infinite Möbius band). Consider on $\mathbb{C}\backslash\{0\}$ the Weierstrass representation given by the form $\omega = f dz$, $f(z) = i(z+1)^2/z^4$ and the meromorphic function $g(z) = z^2(z-1)/(z+1)$ (an example of Meeks, see [10]). Consider on $\mathbb{C}\backslash\{0\}$ the antiholomorphic involution $\widetilde{I}$: $z \to -1/\overline{z}$. This involution is the composition of inversion with respect to the unit circle $\{|z| = 1\}$ and reflection about the origin $z = 0$.

In fact, this Weierstrass representation determines a nonorientable minimal surface (a two-sheeted covering). For

$$g(\widetilde{I}(z)) = g\left(-\frac{1}{\overline{z}}\right) = -\frac{\overline{z}+1}{\overline{z}^2(\overline{z}-1)} = -\frac{1}{\overline{g(z)}}$$

and

$$\widetilde{I}^*\omega = i(\overline{z}+1)^2\overline{z}^2\frac{d\overline{z}}{\overline{z}^2} = i(\overline{z}+1)^2 d\overline{z} = -\overline{g^2\omega}.$$

Therefore the sets $\{|z| \geq 1\}$ and $\{0 < |z| \leq 1\}$ specify the same nonorientable minimal surface $\psi\colon M \to \mathbb{R}^3$ (this surface is immersed, since $f(z) = 0$ only when $z = 1$, but $fg^2(1) = i(z+1)^2|_{z=1} \neq 0$). The resulting surface $\psi\colon M \to \mathbb{R}^3$ is nonorientable, since M is obtained from the set $\{|z| \geq 1\}$ by identifying opposite points of the circle $\{|z| = 1\}$. Topologically M is a projective plane with a point (at infinity) removed, that is, a Möbius band. This band is twisted once. In fact, if (x^1, x^2, x^3) are the standard coordinates in $\mathbb{R}^3$, and $n = (n_1, n_2, n_3)$ is the unit normal to $\widetilde{\psi}\colon \widetilde{M} \to \mathbb{R}^3$, while $g(z)$ is the stereographic coordinate of $n(z)$, that is, $|z| = 1$, we have

$$|g|^2 = \frac{1 - \operatorname{Re} z}{1 + \operatorname{Re} z}, \qquad n_3 = \frac{|g|^2 - 1}{|g|^2 + 1} = -\operatorname{Re} z,$$

so under a motion round the circle $|z| = 1$ from the point $z = 1$ to the point $z = -1$ the normal is parallel to the plane $x^3 = 0$ once (the semicircle $\{e^{i\varphi}, 0 \leq \varphi < \pi\}$ is identified with the semicircle $\{e^{i\varphi}, -\pi \leq \varphi < 0\}$).

§5. Total curvature and complete minimal surfaces

To understand this section we need to have an idea of the volume form on an orientable Riemannian manifold (the area form in the case of two-dimensional surfaces), and also to be able to integrate p-forms over p-dimensional manifolds (in the case $p = 1$ the manifold is a curve; the definition of the integral in this case was given in §4). We also need to be familiar with the classification of closed (compact without boundary) two-dimensional manifolds. For a detailed acquaintance with these questions we recommend the reader to turn if necessary to [1] and [15].

An important characteristic of minimal surfaces in $\mathbb{R}^3$ is their total curvature. The *total curvature* of an orientable surface is defined as the integral of the Gaussian curvature over the surface, and in the nonorientable case as half the total curvature of the orientable doubling of this surface (see §4).

More formally, let $\psi\colon M \to \mathbb{R}^3$ be an immersed orientable surface, α the area form on M in the metric induced by the immersion ψ, and K the Gaussian curvature.

DEFINITION. The *total curvature* $\tau(M)$ of a surface $\psi\colon M \to \mathbb{R}^3$ is the number $\tau(M) = \int_M K \cdot \alpha$ (possibly equal to infinity).

As we mentioned in the proof of Proposition 1 in §3 of Chapter 3, the area of the image of the coordinate parallelogram $\partial/\partial u$, $\partial/\partial v$ of the surface M

under the Gaussian mapping $n\colon M \to S^2$ is multiplied by K, where K is the Gaussian curvature of M.

REMARK. In §3 of Chapter 3 the coordinates (u, v) were isothermal. In fact, it is easy to see that this does not depend on the coordinates (u, v) and the tangent parallelogram: under a smooth mapping of one two-dimensional Riemannian manifold onto another the area of any parallelogram in the tangent space is multiplied by the same number d. This number is called the *determinant* of the tangent mapping.

EXERCISE. Write down the determinant d in local coordinates.

From what we have said it follows that if β is the area form on S^2, and $n\colon M \to S^2$ is the Gaussian mapping, then the 2-form $n^*\beta$ on M, defined as $(n^*\beta)(\xi, \eta) = \beta(n_*\xi, n_*\eta)$ where $\xi, \eta \in T_pM$, is equal to $K\alpha$, where α is the area form on M. In fact, any 2-form on a two-dimensional manifold is uniquely determined by its value on a pair of linearly independent vectors. If $\xi = \partial/\partial u$ and $\eta = \partial/\partial v$, then

$$\begin{aligned} n^*\beta\left(\frac{\partial}{\partial u}, \frac{\partial}{\partial v}\right) &= \beta\left(n_*\frac{\partial}{\partial u}, n_*\frac{\partial}{\partial v}\right) \\ &= \text{area of parallelogram } \left(n_*\frac{\partial}{\partial u}, n_*\frac{\partial}{\partial v}\right) \\ &= K \times \text{area of parallelogram } \left(\frac{\partial}{\partial u}, \frac{\partial}{\partial v}\right) \\ &= K\cdot\alpha\left(\frac{\partial}{\partial u}, \frac{\partial}{\partial v}\right). \end{aligned}$$

If the Gaussian mapping is locally (in a domain $U \subset M$) a diffeomorphism, then $\int_U K\alpha = \int_U n^*\beta = \pm\int_{n(U)}\beta = \pm$ area of the domain $n(U) \subset S^2$. We put the $+$ sign when the diffeomorphism $n\colon U \to n(U)$ preserves the orientation, and the $-$ sign otherwise. The first possibility is realized when $K > 0$ in U, and the second when $K < 0$ (verify this).

Thus, the total curvature $\tau(M)$ of a surface $\psi\colon M \to \mathbb{R}^3$ is equal to the "algebraic" area of the Gaussian image $n(M)$. If K is positive everywhere or negative everywhere, then the total curvature $\tau(M)$ is equal (up to the sign) to the area of the Gaussian image $n(M)$ taking account of overlappings.

As we have already said, among minimal surfaces in $\mathbb{R}^3$ there is most interest in complete minimal surfaces. It turns out that if a complete minimal surface has finite total curvature, then it has quite a simple topological structure, namely, such a surface is isometric to a compact orientable two-dimensional Riemannian manifold without boundary from which finitely many points have been removed. (An isometry is a diffeomorphism that preserves the metric. Two Riemannian manifolds are isometric if there is an isometry between them.)

All two-dimensional connected closed (compact without boundary) manifolds can be classified quite easily. We recall that from the topological point of view they are: in the orientable case a sphere to which a finite number g of handles have been attached, and in the nonorientable case a sphere to which a finite number h of Möbius bands have been attached. The numbers g and h are called the *genus* of the closed manifold in the orientable and nonorientable cases respectively, and the closed manifold itself is called an *orientable* (*nonorientable*) *surface of genus g* (*genus* h).

A surface of genus 0 is a sphere S^2, a surface of genus $g = 1$ is a torus $T^2 = S^1 \times S^1$, and a surface of genus $g = 2$ is a pretzel. In the unorientable case, a surface of genus $h = 1$ is a projective plane, and a surface of genus $h = 2$ is a Klein bottle (see Figure 39).

If from a surface of genus g or h we remove a finite number l of points (or equivalently small closed disjoint disks), then the number l is called the *connectivity* of this surface: a sphere S^2 with l holes can be represented as an l-connected domain, the projective plane $\mathbb{R}P^2$ with one hole is a Möbius

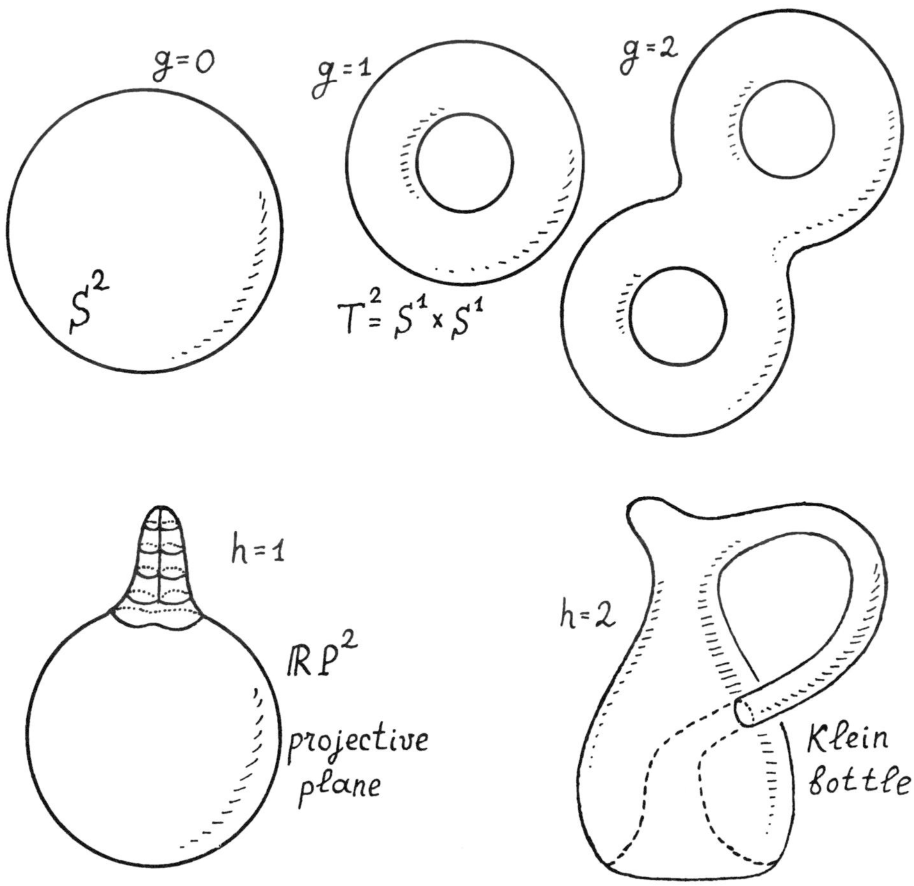

FIGURE 39

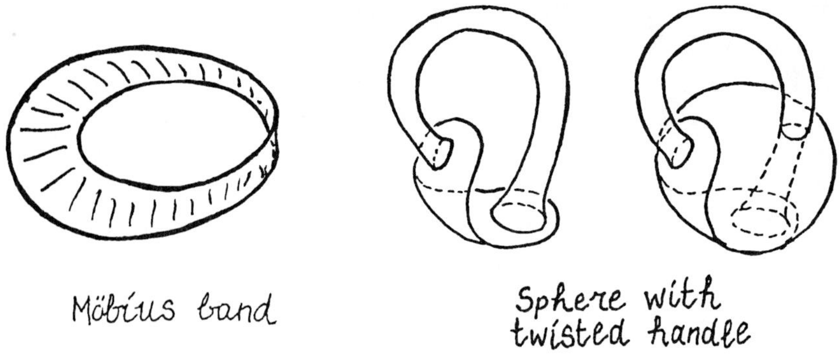

FIGURE 40

band, and a Klein bottle with a hole is called a disk with a twisted handle (see Figure 40).

EXERCISE. Show that a sphere with g handles, h bands, and l holes is diffeomorphic to a sphere with $2g+h$ bands and l holes if $h \geq 1$.

THEOREM (Osserman [9]). *Let* ψ: $M \to \mathbb{R}^3$ *be an immersed complete orientable surface, and suppose that the Gaussian curvature* $K \leq 0$ *and the total curvature* $\tau(M)$ *is finite,* $|\tau(M)| < \infty$. *Then* M *is isometric to a surface* $\overline{M}$ *of genus* g *from which finitely many points* $P_1, \ldots, P_l$ *have been removed. It is assumed that* $\overline{M}$ *is Riemannian, and that on* M *we consider the metric induced by the immersion* ψ.

COROLLARY. *The theorem is true for complete immersed minimal surfaces* ψ: $M \to \mathbb{R}^3$ *of finite total curvature. Moreover, if* (ω, g) *is a Weierstrass representation of such a minimal surface, then the function* g *that gives the Gaussian mapping can be extended at the deleted points* $P_1, \ldots, P_l$ *to a meromorphic function defined on the whole of* $\overline{M}$. *Moreover, the total curvature* $\tau(M)$ *can take only the values* $-4\pi m$, *where* m *is a nonnegative integer.*

REMARK. The condition that the minimal surface is regular is essential. We can construct complete nonplanar branched minimal surfaces whose Gaussian image lies in an arbitrarily small domain of the sphere S^2. For this it is sufficient to construct on the unit disk $D = \{|z| < 1\}$ a Weierstrass representation (f_ε, g) such that $g = \varepsilon z$, $\varepsilon > 0$, and the holomorphic function f_ε can be chosen in such a way that for any curve C: $z(t)$, $t \in (0, 1)$, lying in D and not completely contained in any compactum $K \subset D$ (a divergent curve) we have

$$\int_C |f_\varepsilon(z)|\,|dz| = \int_0^1 |f_\varepsilon(z(t))|\,|\dot{z}(t)|\,dt = \infty.$$

This Weierstrass representation determines a complete branched minimal surface with Gaussian mapping given by the function $g = \varepsilon z$, $|z| < 1$.

REMARK. The function f_ε may have infinitely many zeros in the unit disk D. As Osserman showed, if f has only finitely many zeros in D, then there is always a divergent curve C such that $\int_c |f(z)|\,|dz| < \infty$ (see [9]).

Let us give the idea of the proof of the last two assertions of the corollary. The fact that the function g can be extended at the deleted points to a meromorphic function follows from the classification of isolated singularities of functions holomorphic in a punctured disk (a removable singular point, a pole, an essential singularity) and the great Picard theorem, which asserts that in any neighborhood of an essential singularity a holomorphic function takes any finite value (with one possible exception) infinitely many times. From this we can deduce that if an isolated point P_k is an essential singularity for a function g, then the total curvature $\tau(M)$ is infinite.

The second assertion follows from the formula for transforming the integral of a form under a smooth mapping, namely, if $f\colon M_1 \to M_2$ is a smooth mapping between two smooth closed (compact without boundary) connected orientable two-dimensional manifolds, and α is an arbitrary 2-form on M_2, then $\int_{M_1} f^*\alpha = \deg f \cdot \int_{M_2} \alpha$ where the integer $\deg f$ is called the *degree* of the mapping f and defined as follows.

Consider an arbitrary regular point $Q \in M_2$ for the mapping f, that is, a point such that the inverse image $f^{-1}(Q)$ of Q consists of points $\{P_k\}$ at each of which the mapping f is regular. This means that the tangent mapping $f_*\colon T_{P_k}M_1 \to T_QM_2$ is nonsingular for each point P_k. From the fact that M_1 is compact it follows that $f^{-1}(Q)$ consists of finitely many points $P_1, \dots, P_r$ (these points are isolated).

Let us choose oriented atlases on M_1 and M_2 and let (u_k, v_k) be coordinates on M_1 in a neighborhood of P_k, and (u, v) coordinates on M_2 in a neighborhood of Q. Then

$$\deg f = \sum_{k=1}^{r} \operatorname{sign}\det \begin{pmatrix} \frac{\partial u}{\partial u_k} & \frac{\partial u}{\partial v_k} \\ \frac{\partial v}{\partial u_k} & \frac{\partial v}{\partial v_k} \end{pmatrix}.$$

We can show that a regular point always exists (the set of nonregular points on M_2 has measure zero—Sard's lemma) and that $\deg f$ does not depend on the choice of regular point Q (see [1]). If we change the orientation on one of the manifolds M_1 or M_2, then $\deg f$ changes sign.

For f we choose the extension $\overline{n}$ of the Gaussian mapping $n\colon M \to S^2$ to a smooth mapping $\overline{n}\colon \overline{M} \to S^2$. Since M is oriented with respect to the normal n, which is also the outward normal to the sphere S^2, in our case the signs of the Jacobians from the definition of the degree of a mapping are the same and equal to -1. Therefore $\deg \overline{n} = -m$, where m is the number of inverse images of any point $Q \in S^2$ that is regular with respect to $\overline{n}$. Since $\tau(M) = \int_M n^*\alpha = \int_{\overline{M}} \overline{n}^*\alpha$, where α is the area form on S^2, and $\int_{\overline{M}} \overline{n}^*\alpha = \deg \overline{n} \int_{S^2} \alpha = -4\pi m$, we have $\tau(M) = -4\pi m$, which proves the last assertion.

Now let M be a nonorientable immersed manifold, $\widetilde{M}$ its orientable doubling, and $\pi\colon \widetilde{M} \to M$ the corresponding projection, a two-sheeted covering. It is natural to define the total curvature $\tau(M)$ as half the total curvature $\tau(\widetilde{M})$. Thus the total curvature of a complete nonorientable immersed minimal surface can take only the values $\tau(M) = -2\pi m$, where m is a nonnegative integer.

Let us consider some examples.

1) The plane is a complete minimal simply-connected surface of genus 0 with total curvature $\tau = 0$.

2) The Enneper surface is a complete minimal simply-connected surface of genus 0 with total curvature $\tau = -4\pi$, since the function g of the Weierstrass representation is $g(w) = w$ and is defined on the whole plane $\mathbb{C}$, so the Gaussian mapping is a diffeomorphism of the Enneper surface onto a sphere without the North Pole $S^2 \backslash N$.

3) The catenoid is a complete minimal doubly-connected surface of genus 0 with total curvature $\tau = -4\pi$.

4) The helicoid is a complete minimal simply-connected surface of genus 0 with total curvature $\tau = -\infty$ (the function $g = e^{-\omega}$ specifies an infinite-sheeted covering of the helicoid onto the sphere without the North and South Poles).

5) The incomplete Scherk surface $z = \ln(\cos x / \cos y)$ over $\{|x| < \pi/2, |y| < \pi/2\}$ has Weierstrass representation $(4/(1-w^4), w)$ on the open disk $\{|w| < 1\} \subset \mathbb{C}$. The image of the Gaussian mapping is an open hemisphere, so the total curvature $\tau = -2\pi$. The complete Scherk surface obviously has total curvature $\tau = -\infty$.

6) All periodic complete minimal surfaces (the Schwarz-Riemann surface, the gyroid, and so on) have total curvature $\tau = -\infty$.

These are examples of orientable minimal surfaces.

An example of an unorientable minimal surface is

7) an immersion of a Möbius band specified on $\mathbb{C} \backslash \{0\}$ by the Weierstrass representation $(i(w+1)^2/w^4, w^2(w-1)/(w+1))$ (an example of Meeks, see §4).

The Gaussian mapping is given by the function $g(w) = w^2(w-1)/(w+1)$. If $w^2(w-1)/(w+1) = \eta$, then the equation $w^2(w-1) - \eta(w+1) = 0$ has three roots for each η. Two roots coincide if the derivative of this polynomial with respect to w also vanishes, that is, if $\eta = 3w^2 - 2w$. Substituting this into the first equation, we obtain $-2w(w^2 + w - 1) = 0$, which has three roots $\{w_k\}_{k=1}^3$. The corresponding values $\eta_k = g(w_k)$ are the singular values, and all the remaining $w \in \mathbb{C} \backslash \{\eta_1, \eta_2, \eta_3\}$ are regular. Since the inverse image of a regular value consists of exactly three points, the total curvature of this surface is equal to $\tau = -12\pi/2 = -6\pi$.

In fact, the Möbius band can be immersed in $\mathbb{R}^3$ with total curvature $\tau = -2\pi$, but then this immersion is branched, that is, there are branch points. The branched immersion of an infinite Möbius band as a complete minimal surface has been known for quite a long time—it is the classical Henneberg surface. The Weierstrass representation that specifies the oriented doubling of the Henneberg surface is $(2(1-1/w^4)dw, w)$ on $\mathbb{C}\setminus\{0\}$, and the branch points are the fourth roots of unity, that is, $w = \pm i, \pm 1$.

We note that for branched minimal surfaces the Gaussian mapping is also well defined, since the function g can be extended to a meromorphic function at a branch point. The completeness of the induced metric is also well defined (despite the fact that the metric is degenerate at branch points). In fact, we can define the completeness of an immersed surface in the following equivalent way (see [9]). We call an immersed surface $x\colon M \to \mathbb{R}^3$ *complete* if any divergent curve on M, that is, a ray $c\colon [0, +\infty) \to M$ that does not lie entirely in one compact subset $K \subset M$ has infinite length, that is, $\int_0^\infty |\dot{c}(t)|\,dt = \infty$. Here $|\dot{c}(t)|$ is the length of the velocity vector of this curve in the metric induced from $\mathbb{R}^3$. Since the integral $\int_0^\infty |\dot{c}(t)|\,dt$ is defined for the metric, which is degenerate at certain (isolated) points, a similar definition of completeness carries over to branched surfaces.

It turns out that there are not many complete immersed minimal surfaces with total curvature greater than -8π. Namely, we have the following result.

THEOREM (Osserman [9], Meeks [10]). *The complete immersed connected minimal surfaces in $\mathbb{R}^3$ with total curvature $\tau > -8\pi$ are exhausted* (*up to a motion and an expansion in $\mathbb{R}^3$*) *by the following list*:

the plane ($\tau = 0$),
the catenoid ($\tau = -4\pi$), *the Enneper surface* ($\tau = -4\pi$),
the Möbius band (*Meeks's example,* $\tau = -6\pi$).

REMARK. The total curvature of the complete Henneberg surface is equal to $\tau = -2\pi$, but this surface is branched. If we delete the branch points, the Henneberg surface becomes incomplete.

From Osserman's theorem, which describes complete minimal surfaces with finite total curvature, the following natural questions arise:

1) Can any of the possible values of the total curvature be realized on some complete minimal surface?

2) Are there complete minimal surfaces of any genus and any connectivity?

The answer to the first question is given by the following theorem.

THEOREM. *For any integer $k \geq 0$, except $k = 1$, there is a complete immersed minimal surface $\psi\colon M \to \mathbb{R}^3$ with total curvature $\tau(M) = -2\pi k$. The case $k = 1$ cannot be realized.*

PROOF. For each even $k = 2m$ we can easily construct a complete orientable immersed minimal surface with total curvature $\tau(M) = -4\pi m$. Each such surface can be specified, for example, by a Weierstrass representation (dw, w^m) on $\mathbb{C}$.

The fact that the case $k = 1$ cannot be realized follows from the theorem of Osserman and Meeks formulated above.

It remains to construct examples for each odd $k \geq 3$. This problem was solved by Elisa and d'Oliveira (see [11]), and also by Kusner (see [12]). We now give the examples of the first two authors, and we shall talk about the somewhat more interesting Kusner surfaces in the next section. Thus, Elisa and d'Oliveira constructed a family of complete minimal immersions of the infinite Möbius band with total curvature $\tau = -2\pi m$, $m \geq 3$, m odd. The authors also showed that other values of the total curvature cannot occur for complete unorientable minimal surfaces of genus 1. Each surface of this family is specified on $\mathbb{C}\backslash\{0\}$ by the Weierstrass representation

$$\left(\frac{i(w+1)^2}{w^{m+1}}\,dw\,,\ \frac{w^{m-1}(w-1)}{w+1}\right),$$

m odd, $m \geq 3$, and I: $w \to -1/\overline{w}$ is the standard involution on $\mathbb{C}\backslash\{0\}$ without fixed points. Each such surface is a Möbius band with $(m-1)/2$ twists immersed in $\mathbb{R}^3$ as a complete minimal surface with total curvature $\tau = -2\pi m$ (verify this).

The answer to the second question is given in the orientable case by a paper of Klotz and Sario (see [13]), in which they construct complete immersed orientable minimal surfaces of any genus and any connectivity $c \geq 4$.

The idea of the construction is as follows. We first consider a 3-connected complete minimal surface specified by the Weierstrass representation

$$(dw\,,\ g = 1/(w-1) + 2/(w-1)^2 + 1/(w+1) - 2/(w+1)^2)$$

on $\mathbb{C}\backslash\{1, -1\}$. It is easy to verify that the residues of the forms $\varphi^1 = \frac{1}{2}(1-g^2)dw)$ and $\varphi^2 = \frac{i}{2}(1+g^2)\,dw$ at the points $w = \pm 1$ are zero, and the residues of the form $\varphi^3 = g\,dw$ are real numbers, so $\int \varphi^3$ along a closed contour round each singular point is equal to $2\pi i \operatorname{res}_{w=\pm 1}\varphi^3$ (where $\operatorname{res}_{w=\pm 1}\varphi^3$ denotes the residue at the corresponding point), that is, it is a purely imaginary number, and so φ^3 also does not have real periods. Moreover, the induced metric is $d\overline{s} = (1+|g|^2)|dw|$, so any divergent curve going off to infinity has infinite length, since $d\overline{s} \geq |dw|$. Any divergent curve going to a singular point $w = \pm 1$ also has infinite length, since in a neighborhood of these singular points the metric $d\overline{s} \sim |dw|/|1 \pm w|^4$ increases quite rapidly.

Thus, we have constructed a complete immersed 3-connected minimal surface of genus 0.

Next, we take two copies F_1 and F_2 of the plane $\mathbb{C}\backslash\{1, -1\}$ and on each of them we make a cut γ_k, $k = 1, 2$, along the real axis from -1 to 1. We glue the upper side of γ_1, to the lower side of γ_2, and the lower side of γ_1 to the upper side of γ_2. We have obtained a manifold F glued together from F_1 and F_2. It is easy to define a complex structure on F induced by the complex structures on F_1 and F_2, and also a holomorphic form $\widetilde{\omega}$ and a meromorphic function $\widetilde{g}$, which in the coordinates on F induced by the standard coordinates on the corresponding copies F_k coincide with the form dw and the function g defined above. It is easy to verify that the pair $(\widetilde{\omega}, \widetilde{g})$ on F is a global Weierstrass representation, which determines a complete immersed minimal surface of genus 0, but this time 4-connected (see Figure 41).

Continuing this procedure, we can obtain surfaces of genus 0 and any connectivity. To obtain the complete picture for surfaces of genus 0 we add to this family the simply-connected Enneper surface and the doubly-connected catenoid. Thus, there are complete immersed orientable minimal surfaces of genus 0 and any connectivity $c \geq 1$.

To obtain a surface of positive genus we take two copies F^k, $k = 1, 2$, of the 4-connected Riemann surface F defined above (F^k is glued together from F_1^k, and F_2^k, $F_1^k = \mathbb{C}\backslash\{1, -1\}$). We make cuts δ_k and ε_k on each F^k: on F_1^k a cut δ_k from 1 to $+\infty$ along the real axis, and on F_2^k a cut ε_k from -1 to $-\infty$ also along the real axis. We glue together the upper (lower) sides of the cuts δ_1 and ε_1 on F^1 to the lower (upper) sides of the cuts δ_2 and ε_2 on F^2. As a result we obtain a torus with four singular points: we denote it by M_g^c, $g = 1$, $c = 4$ (see Figure 42).

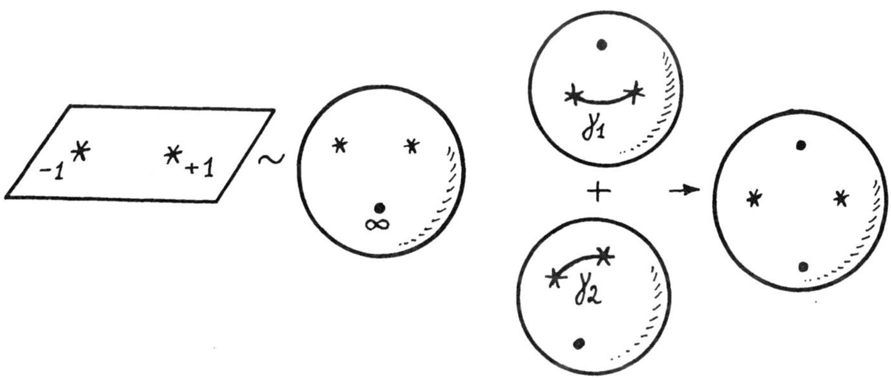

FIGURE 41A

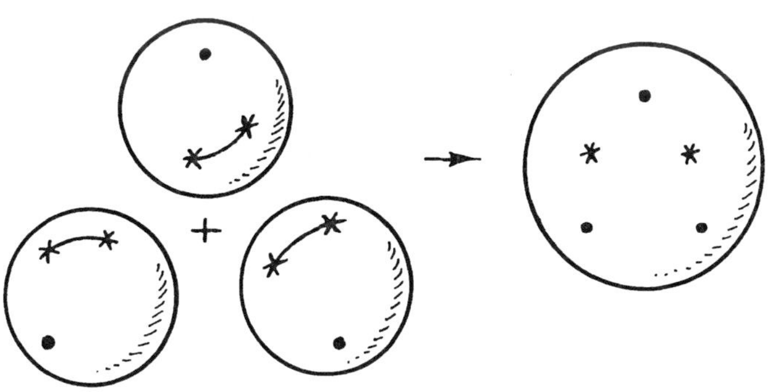

FIGURE 41B

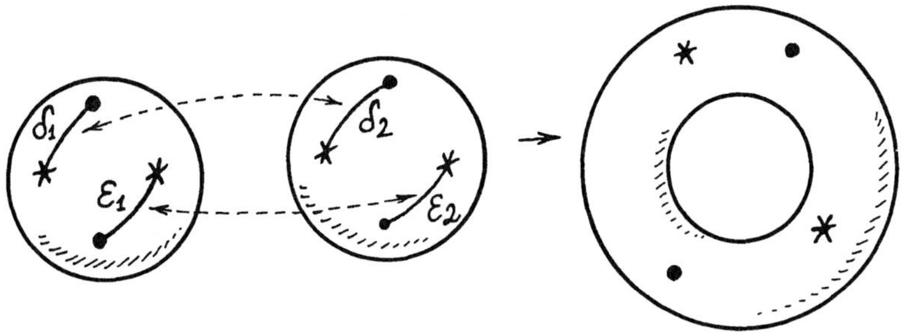

FIGURE 42

As above, we construct the global Weierstrass representation on M_g^c induced by the Weierstrass representations on each of the components from which M_1^4 is glued. We have thus constructed a complete immersed 4-connected minimal surface of genus 1.

Continuing this construction, we can obtain a 4-connected minimal surface M_g^4 of any genus $g \geq 0$. To increase the connectivity (without changing the genus) we make a cut on M_g^4 between the singular points corresponding to $w = \pm 1$ and glue the necessary number of copies of $F_k = C\backslash\{1, -1\}$ (see Figure 43).

IMPORTANT REMARK. We note that the image in $\mathbb{R}^3$ of the minimal surfaces we have constructed (of any genus and any connectivity) coincides as a set of points with the 3-connected minimal surface defined at the very beginning by the Weierstrass representation (dw, g) on $\mathbb{C}\backslash\{1, -1\}$. Of course, this admits an immersion (compare with the infinite-sheeted covering of the catenoid in §2 of Chapter 2). Thus we have simply increased the parametric domain without changing the actual surface as a subset of $\mathbb{R}^3$.

§6. The geometry of complete minimal surfaces of finite total curvature

In this section we describe the examples of Meeks and Hoffman of complete minimal surfaces of arbitrary genus embedded in $\mathbb{R}^3$. All these surfaces have finite total curvature. We first introduce some concepts that characterize the geometry of complete minimal surfaces of finite total curvature that are used in the construction.

We recall that by Osserman's theorem every orientable complete minimal surface $\psi\colon M \to \mathbb{R}^3$ of finite total curvature is isometric to a surface $\overline{M}$ of some genus g from which a finite number r of points $P_1, \ldots, P_r$ have been deleted. It turns out that in a neighborhood of the deleted points $P_1, \ldots, P_r$ the immersion $\psi\colon \overline{M}\backslash\{P_1, \ldots, P_r\} \to \mathbb{R}^3$ behaves quite well (here we again denote by ψ the composition of the isometry between $\overline{M}\backslash\{P_1, \ldots, P_r\}$ and M and the immersion $\psi\colon M \to \mathbb{R}^3$). Namely, let us choose an arbitrary

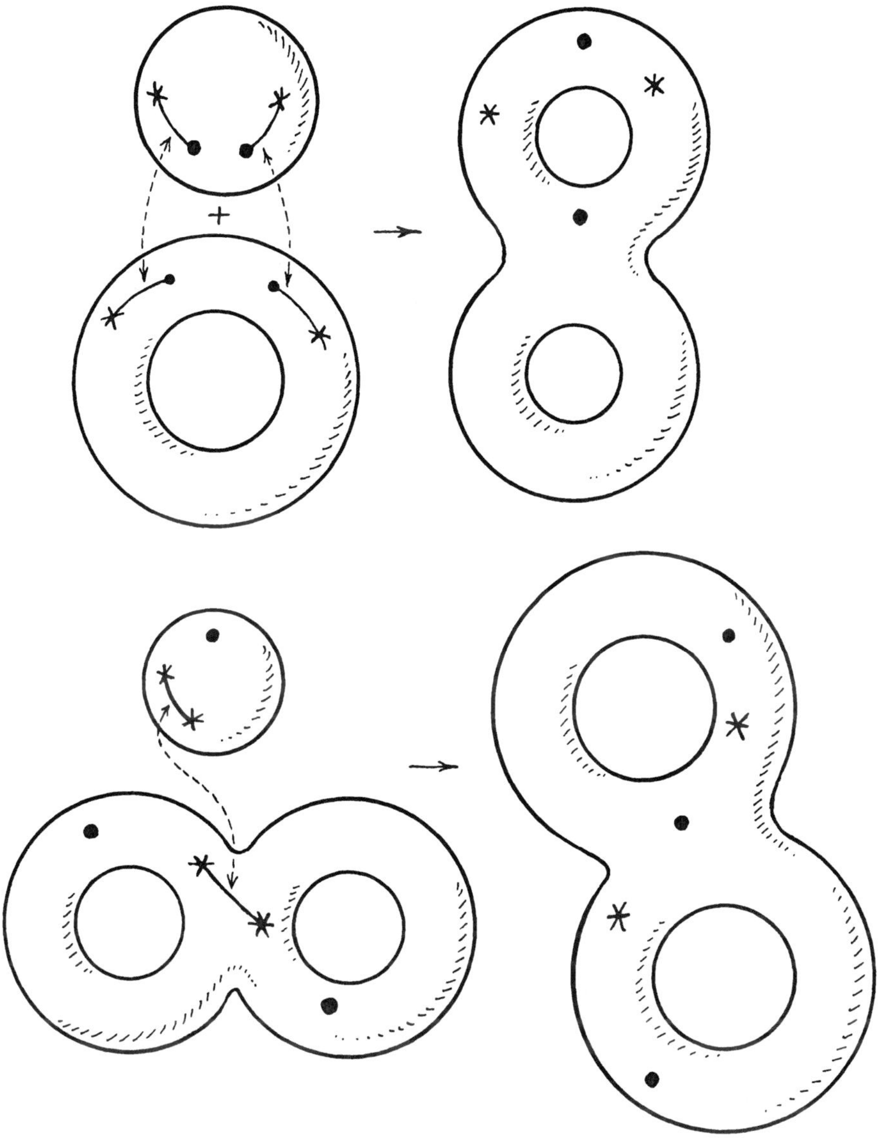

FIGURE 43

$\rho > 0$. Let X_ρ denote the intersection of the image $\psi(M)$ contracted by a factor ρ, that is, $\psi(M)/\rho$, and the standard unit sphere $S^2 \subset \mathbb{R}^3$: $X_\rho = \psi(M)/\rho \cap S^2$.

As Jorge and Meeks showed (see [14]), for sufficiently large ρ the set X_ρ is a family $\{\gamma_1, \ldots, \gamma_r\}$ of smooth closed curves immersed in the sphere S^2.

DEFINTION. We define an *end* of a complete minimal surface $\psi: M \to \mathbb{R}^3$ of finite total curvature as the image $c_l = \psi(D_l \backslash P_l)$ of a sufficiently small punctured disk $D_l \backslash P_l$ on $\overline{M}$ with center at the point P_l.

For sufficiently large ρ the curve γ_l of the family $\{\gamma_1, \dots, \gamma_r\}$ constructed above is the intersection of c_l/ρ and S^2: $\gamma_l = c_l/\rho \cap S^2$. Thus, the asymptotic behavior of the curves characterizes the behavior of the ends of the minimal surface.

It turns out (see [14]) that as $\rho \to \infty$ each curve $\gamma_l \subset S^2$ contracts to a great circle ν_l in S^2 (ν_l is defined as the intersection of S^2 and a plane passing through the origin), and the limit curve $\widetilde{\gamma}_l$ is a winding of ν_l with a certain multiplicity, which we denote by d_l and call the *multiplicity* of the end c_l. Moreover, ν_l lies in the plane perpendicular to the image of the point P_l under the Gaussian mapping $\overline{n}$: $\overline{M} \to S^2$ (an extension of the Gaussian mapping n: $M \to S^2$, which exists by Osserman's theorem). Geometrically this means that each end looks asymptotically like a d_l-sheeted covering of the limiting tangent plane (the plane perpendicular to the "limiting value" of the Gaussian mapping—the image of P_l) onto itself (as in the case of the holomorphic mapping $\mathbb{C} \to \mathbb{C}$ given by the formula $w = z_l^d$).

EXAMPLE. Consider the catenoid $r = \cosh z$ described in cylindrical coordinates (r, φ, z). The catenoid has two ends and under an increase in the scale they are as shown in Figure 44.

Clearly, X_ρ consists of two simple Jordan curves γ_1 and γ_2 lying in parallel planes, and as $\rho \to \infty$ these curves contract to the equator. In this case $d_1 = d_2 = 1$.

PROBLEM 1. Show that the Enneper surface X_ρ for sufficiently large ρ consists of one curve immersed in S^2, which in the limit winds onto the equator with multiplicity $d_1 = 3$.

PROBLEM 2. Let (f, g) on $U \subset \mathbb{C}$ be a local Weierstrass representation of a minimal surface ψ: $M \to \mathbb{R}^3$, and let τ: $\mathbb{R}^3 \to \mathbb{R}^3$ be an orthogonal transformation that preserves the orientation. The surface $\tau \circ \psi$: $M \to \mathbb{R}^3$ is obviously a minimal surface congruent to the surface ψ: $M \to \mathbb{R}^3$ and has a different Weierstrass representation (f', g') over $U \in \mathbb{C}$. Find the connection between (f', g') and (f, g).

ANSWER. If the orthogonal transformation τ: $\mathbb{R}^3 \to \mathbb{R}^3$, restricted to the sphere S^2, is given in stereographic coordinates (u, v), $z = u + iv$,

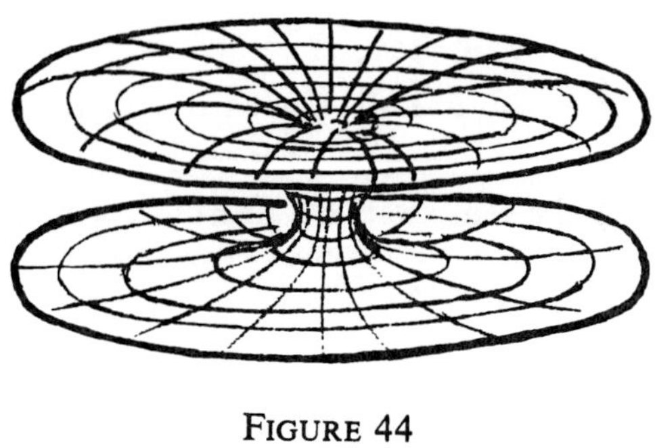

FIGURE 44

by a special unitary matrix $\left(\begin{smallmatrix} a & b \\ -\bar{b} & \bar{a}\end{smallmatrix}\right)$, $a, b \in \mathbb{C}$, $|a|^2 + |b|^2 = 1$ as follows: $z \mapsto (az+b)/(-\bar{b}z+\bar{a})$, then the new Weierstrass representation (f', g') has the form

$$(f', g') = \left((-\bar{b}g+\bar{a})^2 f,\ \frac{ag+b}{-\bar{b}g+\bar{a}}\right).$$

Problem 3. Let (f, g) over U be the Weierstrass representation of a minimal surface $\psi\colon M \to \mathbb{R}^3$, and $\lambda\colon \mathbb{R}^3 \to \mathbb{R}^3$ an expansion by a factor $\lambda \in \mathbb{R}$: $x \mapsto \lambda x$ for any $x \in \mathbb{R}^3$. Prove that the Weierstrass representation (f', g') over U of the minimal surface $\lambda \circ \psi\colon M \to \mathbb{R}^3$ is $(\lambda f, g)$.

Problem 4. Let $(f(z), g(z))$ over U be the Weierstrass representation of a minimal surface $\psi\colon M \to \mathbb{R}^3$, let $\alpha\colon U \to V$ be a holomorphic change of coordinates, and let w be the standard complex coordinate in V. Prove that the Weierstrass representation (f', g') of the minimal surface $\psi\colon M \to \mathbb{R}^3$ over V in the new coordinates has the form $(\partial z/\partial w \cdot f, g)$.

The key feature in the proof of the above assertions that describe the asymptotic behavior of the ends of a complete minimal surface of finite total curvature is the following remark: if (ω, g) is the global Weierstrass representation of a complete minimal surface $\psi\colon M \to \mathbb{R}^3$ of finite total curvature, $M \approx \overline{M}\backslash\{P_1, \dots, P_r\}$, then not only can the function g be extended to a meromorphic function on $\overline{M}$, but also the holomorphic 1-form ω can be extended on $\overline{M}$ to a meromorphic 1-form. In fact, if this is true for the Weierstrass representation (ω, g) of a minimal surface $\psi\colon M \to \mathbb{R}^3$, then it is also true for the Weierstrass representation (ω', g') of the minimal surface $\tau \circ \psi\colon M \to \mathbb{R}^3$, where $\tau\colon \mathbb{R}^3 \to \mathbb{R}^3$ is an arbitrary orthogonal transformation that preserves the orientation (see Problem 2). Obviously, for each point $P_l \in \{P_1, \dots, P_r\}$ we can select an orthogonal transformation τ_l such that P_l becomes a zero of the function $\tilde{g}'$, the extension of the function g' from the Weierstrass representation for $\tau_l \circ \psi\colon M \to \mathbb{R}^3$. Since this minimal surface is also complete, any smooth (divergent) curve γ that tends to P_l has infinite length, that is, $\int_\gamma |f'|(1+|g'|^2)\,dt = \infty$. Since $g'(P) \to 0$ as $P \to P_l$, we have $\int_\gamma |f'|\,dt = \infty$.

Hence it follows (try to show this) that $|f'(P)| \to \infty$ as $P \to P_l$, that is, f' has a pole at P_l.

Problem 5. Prove that the only (up to a motion and an expansion in $\mathbb{R}^3$) simply-connected and doubly-connected immersed complete minimal surfaces $\psi\colon M \to \mathbb{R}^3$ of genus 0 with total curvature $\tau(M) = -4\pi$ are the Enneper surface and the catenoid respectively.

Hint. A complete simply-connected immersed minimal surface of genus 0 and finite total curvature is given by the global Weierstrass representation $(\omega = f\,dz, g)$ on the plane $\mathbb{C}$, where g and f are extended to meromorphic functions on $S^2 \approx \mathbb{C} \cup \{\infty\}$, and f is a holomorphic function on $\mathbb{C}$. Therefore f is a polynomial, and $g = P/Q$ is a complex function, where P and Q are polynomials, and because the surface is immersed $f = cQ^2$, where $c = \text{const}$. Thus, all such surfaces are described by the Weierstrass representation $(cQ^2, P/Q)$ on $\mathbb{C}$, where P and Q are coprime polynomials, and the total curvature is calculated from the formula $\tau(M) = -4\pi \max(\deg P, \deg Q)$, where deg denotes the degree of a polynomial. Similar arguments can be carried out for a doubly-connected surface.

In the investigation of minimal surfaces for embeddedness an important role is played by a generalization of the classical Gauss-Bonnet theorem. It turns out that we can calculate the finite total curvature of a complete minimal surface $\psi: M \to \mathbb{R}^3$ by starting from the topology of the manifold $\overline{M}$ and the geometry of ends. If g is the genus of $\overline{M}$, $M \approx \overline{M}\backslash\{P_1, \dots, P_r\}$, $d_1, \dots, d_r$ are the multiplicities of the ends $c_1, \dots, c_r$, and $\tau(M)$ is the total curvature, then we have the following result.

THEOREM 1 (a generalization of the Gauss-Bonnet formula; see [9], [14]).

$$\tau(M) = 2\pi\left(2 - 2g - r - \sum_{j=1}^{r} d_j\right).$$

The quantity $2 - 2g$ *is called the Euler characteristic of* $\overline{M}$ *and denoted by* $\chi(\overline{M})$. *Thus,*

$$\tau(M) = 2\pi\left(\chi(\overline{M}) - r - \sum_{j=1}^{r} d_j\right).$$

EXAMPLES. 1. The catenoid $\psi: M \to \mathbb{R}^3$ has two ends, embedded in $\mathbb{R}^3$, that is, $r = 2$, $d_1 = d_2 = 1$, and $\overline{M} = S^2$, so the Euler characteristic $\chi(\overline{M}) = \chi(S^2) = 2$. Consequently, $\tau(M) = 2\pi(2 - 2 - 2) = -4\pi$, which agrees with the previous results.

2. The Enneper surface $\psi: M \to \mathbb{R}^3$ has one end of multiplicity d_1, $\chi(\overline{M}) = 2$, and the total curvature $\tau(M) = -4\pi$. Therefore $-4\pi = 2\pi(2 - 1 - d_1)$, from which it follows that $d_1 = 3$ (see also above).

From Theorem 1 there follows a necessary condition for a complete minimal surface $\psi: M \to \mathbb{R}^3$ of finite total curvature $\tau(M)$ to be embedded. Since the ends of such a surface must be embedded, which is obviously equivalent to the multiplicity of any end being equal to one, the generalized Gauss-Bonnet formula leads to the relation $\tau(M) = 2\pi(\chi(\overline{M}) - 2r)$ or $g + r - 1 = -\tau(M)/4\pi =$ the degree of the Gaussian mapping.

Moreover, all the ends must be "parallel", that is, the Gaussian mapping of the surface $\overline{M}$ must take all the points $P_1, \dots, P_r$ into a pair of opposite points of the sphere S^2 (otherwise for sufficiently large ρ the curves of the family $X_\rho = \{\gamma_1, \dots, \gamma_r\}$ will intersect and the surface will not be embedded). After a certain rotation we can arrange that $P_1, \dots, P_r$ go into the North and South Poles. If (ω, g) is the global Weierstrass representation of such a surface, and $\tilde{g}$ is the extension of g to $\overline{M}$, then the last condition means that $P_1, \ldots, P_r$ are either zeros or poles of $\tilde{g}$.

SKETCH OF THE PROOF OF THEOREM 1. First of all we note that the Euler characteristic of a closed two-dimensional surface of genus g has an intuitive geometrical definition. To start with we consider a closed convex polyhedron. Euler's well-known theorem asserts that the number of faces f, the number of edges e, and the number of vertices v of any such polyhedron satisfy the relation $f - e + v = 2$. Let us project the polyhedron by means of a central projection onto an arbitrary sphere lying inside it. We obtain a partition of this sphere into domains corresponding to the faces of the polyhedron, and the boundary of each domain obviously splits into "edges" and "vertices". The relation $f - e + v = 2$ naturally remains true for such a partition of the sphere.

In addition, let us consider an arbitrary closed surface M of genus g. We split M into finitely many disjoint polygonal domains. This means that we represent M as the union of the closures $\overline{U}_i$ of finitely many disjoint open domains U_i (faces), each of which is homeomorphic by means of a coordinate homeomorphism to a simply-connected domain, and the boundary ∂U_i of each face U_i is representable as finitely many edges—smooth curves homeomorphic to an interval, and finitely many points—vertices. We require that each edge adjoins (that is, lies in the closure of) at most two domains. The figure shows a possible partition of the torus.

Let f denote the number of faces, e the number of edges, and v the number of vertices. Then it turns out that the quantity $\chi(M) \overset{\text{def}}{=} f - e + v$ does not depend on the partition of M into polygonal domains. It is easy to calculate $\chi(M)$ by representing M as the result of glueing a $4g$-gon (see §5). The corresponding partition consists of one face—the polygon itself, one vertex, and $2g$ edges. Consequently, $\chi(M) = 2 - 2g$.

This result is called the *global Gauss-Bonnet theorem* and can be obtained by calculating the total curvature.

THEOREM 2 (the global Gauss-Bonnet theorem). *Let $\psi\colon M \to \mathbb{R}^3$ be an immersion of a closed connected surface M of genus g in $\mathbb{R}^3$. Then for any partition of M into polygonal domains the total curvature $\tau(M)$ is equal to $\tau(M) = \int_M K = 2\pi(f - e + v) \overset{\text{def}}{=} 2\pi\chi(M)$, where f is the number of faces, e is the number of edges, and v is the number of vertices of the partition, and K is the Gaussian curvature of the metric induced by the immersion ψ.*

Thus, the global Gauss-Bonnet theorem proves that the quantity $\chi(M) = f - e + v$ is independent of the partition, and the total curvature $\tau(M)$ is independent of the immersion $\psi\colon M \to \mathbb{R}^3$.

To derive a relation between the total curvature $\int_M K$ and the Euler characteristic $\chi(M)$ it is sufficient to calculate the total curvature of an arbitrary simply-connected polygonal domain on M. The answer is given by the so-called *local Gauss-Bonnet theorem.*

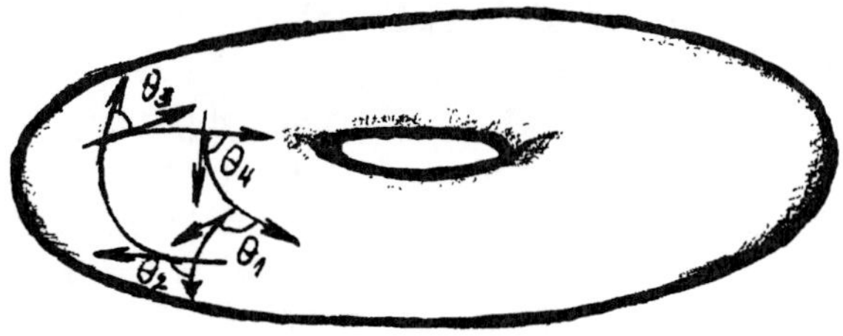

FIGURE 45

THEOREM 3 (the local Gauss-Bonnet theorem). *Let U be a simply-connected polygonal domain on an immersed surface ψ: $M \to \mathbb{R}^3$, let $\theta_1, \dots, \theta_k$ be the exterior angles of U, defined in the obvious way between the tangent vectors to the edges at the vertices of U (see Figure 45), and let k_g be the geodesic curvature of the edges of the boundary ∂U of U. Then*

$$\int_U K + \int_{\partial U} k_g = 2\pi - \sum_{l=1}^{k} \theta_l.$$

Let us recall the definition of the geodesic curvature of a curve on a surface. Let $\gamma(s)$ be a curve on a surface ψ: $M \to \mathbb{R}^3$, parametrized by arc length. This means that $|\dot{\gamma}(s)| = 1$. Hence it follows that the acceleration $\ddot{\gamma}(s)$ is perpendicular to the velocity $\dot{\gamma}(s)$. Let us orient M, and let $J\dot{\gamma}(s)$ denote the unit vector tangent to M and perpendicular to $\dot{\gamma}(s)$; the pair $\dot{\gamma}(s)$, $J\dot{\gamma}(s)$ forms a positively oriented frame of the tangent space $T_{\gamma(s)}M$. Then the geodesic curvature $k_g(s)$ of the curve $\gamma(s)$ is defined as the scalar product $\langle \ddot{\gamma}(s), J\dot{\gamma}(s) \rangle$, which up to the sign is equal to the length of the orthogonal projection of the acceleration $\ddot{\gamma}(s)$ on the tangent plane $T_{\gamma(s)}M$, that is, the tangential component of the acceleration $\ddot{\gamma}(s)$. The curve $\gamma(s)$ is called a geodesic if its geodesic curvature is zero: $k_g(s) = 0$. Thus, the geodesic curvature measures the "deviation" of a curve lying on a surface from the geodesic. It is easy to see that the differential equation that gives the geodesics in some local coordinates on M is an ordinary second-order differential equation, so from the theorem on the existence and uniqueness of solutions it follows that from any point of the surface and in any direction we can draw a unique (up to parametrization) geodesic. For sufficiently close points P and Q of a complete surface M there is always a geodesic γ joining them, and among all piecewise-smooth curves joining P and Q the geodesic γ has the shortest length. Thus, geodesics are the extremals of the length functional, defined on the space of all piecewise-smooth curves joining a pair of points P and Q of the surface M (compare with §1 of Chapter 1). If we consider the area functional as the generalization of the length functional, then minimal surfaces, which are the extremals of the area functional, can be regarded as a generalization of geodesics (for a more general definition of geodesics and a description of their properties, see [1]).

The global Gauss-Bonnet theorem follows from the local theorem in the following way. For an arbitrary partition $\{U_i\}$ of an immersed surface $\psi\colon M \to \mathbb{R}^3$ of genus g into polygonal domains we calculate the total curvature $\tau(M)$ as the sum of the total curvatures of the faces U_i. For each face we use the local Gauss-Bonnet theorem

$$\int_{U_i} K + \int_{\partial U_i} k_g = 2\pi - \sum \theta_l.$$

After summation over all the faces the integrals of the geodesic curvatures cancel out, since we pass through each edge l exactly twice in opposite directions, and the integrals $\int_{l\pm} k_g$ in the two cases are equal in modulus and opposite in sign (verify this).

Next, if $\alpha_l = \pi - \theta_l$ are the corresponding internal angles of the polygonal domains, then on the right we obtain the sum

$$\begin{aligned}\sum_{\text{faces}} \left(2\pi - \sum_{\substack{\text{vertices}\\ \text{of a face}}} \theta_e \right) &= \sum_{\text{faces}} 2\pi - \sum_{\text{faces}} \sum_{\substack{\text{vertices}\\ \text{of a face}}} (\pi - \alpha_l) \\ &= \sum_{\text{faces}} 2\pi - \sum_{\text{faces}} \sum_{\substack{\text{vertices}\\ \text{of a face}}} \pi + \sum_{\text{faces}} \sum_{\substack{\text{vertices}\\ \text{of a face}}} \alpha_l.\end{aligned}$$

The first sum is equal to $2\pi f$. The second sum is equal to $2\pi e$, since if we fix a vertex then the number of faces meeting at this vertex is equal to the number of edges meeting there. When we sum over all the vertices, each edge will be counted twice. The third sum is equal to $2\pi v$, since the sum of the interior angles at any vertex is 2π. It remains to sum over the vertices. This completes the proof of the global Gauss-Bonnet theorem.

We shall not give a proof of the local Gauss-Bonnet theorem. A reader who is interested can refer to [15], for example.

We observe that the total curvature of a polygonal domain has the following geometrical meaning. Let U be an n-gonal simply-connected domain on a surface $\psi\colon M \to \mathbb{R}^3$, where the boundary ∂U consists of geodesic segments (called a geodesic n-gon). Then by the local Gauss-Bonnet theorem we have

$$\int_U K = 2\pi - \sum_{l=1}^{n} \theta_l = \pi(2-n) + \sum_{l=1}^{n} \alpha_l$$

that is,

$$\sum_{l=1}^{n} \alpha_l = \pi(n-2) + \int_U K.$$

As we know, for a planar n-gon the sum of the interior angles is equal to $\pi(n-2)$. Thus, the total curvature of a geodesic n-gon is equal to the deviation of the sum of the interior angles of this n-gon from the sum of the interior angles of a planar n-gon.

Remark. In fact, the Gaussian curvature K of an immersed surface $\psi: M \to \mathbb{R}^3$ depends exclusively on the metric induced by the immersion and not on the immersion: if $\psi_k: M \to \mathbb{R}^3$, $k = 1, 2$ are two immersions that induce the same metric on M, then the Gaussian curvature K is the same in the two cases (however, this is not true for the mean curvature). Moreover, we can define the Gaussian curvature of an arbitrary two-dimensional Riemannian manifold (respectively, the total curvature) without using an immersion in $\mathbb{R}^3$. The Gauss-Bonnet theorem remains true (see [1]).

Using the local Gauss-Bonnet theorem, it is not difficult to prove Theorem 1. Let $\psi: M \to \mathbb{R}^3$ be an immersed orientable connected complete minimal surface of finite total curvature, let $\overline{M}$ be the corresponding Riemann surface of genus g, and suppose that M is isometric to $\overline{M}\backslash\{P_1, \dots, P_r\}$. We cut out from $\overline{M}$ a set of sufficiently small disjoint open disks D_k with centers at P_k, $k = 1, \dots, r$. We denote the remaining compact set by B,

$$B = \overline{M}\backslash \bigsqcup_{k=1}^{r} D_k, \qquad \partial B = \bigsqcup_{k=1}^{r} \partial D_k.$$

Consider an arbitrary partition of B into polygonal simply-connected domains; for each of these domains we use the local Gauss-Bonnet theorem. Summing over all the domains of the partition, we obtain

$$\int_B K + \sum_{k=1}^{r} \int_{\partial D_k} k_g = 2\pi(\chi(\overline{M}) - r)$$

(verify this).

If d_k is the multiplicity of the end $c_k = \psi(D_k \backslash P_k)$, then it is not difficult to show that $\int_{\partial D_k} k_g \to 2\pi d_k$ as the disk D_k contracts to the point P_k. Let us give the idea of the proof of this assertion. Suppose that under the Gaussian mapping P_k goes into the South Pole (this can always be achieved by a rotation of $\mathbb{R}^3$). From the local Weierstrass representation (f, g) in a neighborhood of P_k we have

$$\varphi_1 = \frac{a(z)}{z^m}, \quad \varphi_2 = \frac{b(z)}{z^m}, \quad \varphi_3 = \frac{c(z)}{z^{m-l}},$$

where $l, m \geq 0$, and when $z = 0$, which corresponds to the coordinate of the point P_k, the holomorphic functions $a(z)$, $b(z)$, and $c(z)$ do not vanish.

If x^1, x^2, x^3 are the standard coordinates in $\mathbb{R}^3$ and $x^k = \text{const} + 2\,\mathrm{Re} \int \varphi_k \, dz$, then it is easy to show that

$$\rho = \sqrt{(x^1)^2 + (x^2)^2} \to \infty \quad \text{and} \quad \frac{|x^3|}{\sqrt{(x^1)^2 + (x^2)^2}} = \frac{|x^3|}{\rho} \to 0$$

as $z \to 0$. Hence it follows that as $\rho \to \infty$ $\gamma_k = \psi(D_k \backslash P_k)/\rho \cap S^2$ contracts to the unit circle (the equator of the sphere) $(x^1)^2 + (x^2)^2 = 1$. The limiting curve $\tilde{\gamma}_k$ winds onto the equator d_k times, where d_k is the multiplicity of the end c_k.

Let us calculate the limiting values $\int_{\partial D_k} k_g$ as D_k contracts to the point P_k. For sufficiently large ρ the normal to the end c_k tends to the vertical position, and the curve $\gamma_k = \psi(D_k \backslash P_k) \cap S^2$ tends to the horizontal. If a point moves along the curve $\rho\gamma_k = \psi(D_k \backslash P_k) \cap S^2_\rho$ (S^2_ρ is the sphere with center at the origin and radius ρ) with unit speed, then along γ_k in the same parametrization the speed of the point is equal to $1/\rho$. Thus, $\gamma_k \sim (\cos(t/\rho), \sin(t/\rho), 0)$ as $\rho \to \infty$, and t varies from 0 to $2\pi\rho d_k$. Therefore the acceleration vector $\rho\ddot{\gamma}_k$ of the curve $\rho\gamma_k$ is approximately equal to $1/\rho$, and as $\rho \to 0$ it tends to take up a horizontal position. Hence it follows that the length of the tangential component of the vector $\rho\ddot{\gamma}_k$ for sufficiently large ρ is approximately equal to $1/\rho$ and

$$\int_{\rho\gamma_k} k_g \approx \int_0^{2\pi\rho d_k} \frac{1}{\rho}\, dt = 2\pi\, d_k.$$

PROBLEM 6. Prove that the minimal surface given by the Weierstrass representation $((z^{m+1}-1)^{-2}, z^m)$ on $\mathbb{C}\backslash\{z^{m+1}=1\}$ can be extended at $z=\infty$ to a complete minimal surface defined on $S^2\backslash\{z^{m+1}=1\}$, $S^2 = \mathbb{C}\cup\{\infty\}$ and having $m+1$ embedded ends.

Hint. The total curvature of this surface is $-4\pi m$. It remains to use the formula from Theorem 1 and show that the multiplicity of each of the ends is equal to one.

PROBLEM 7. Show that the Enneper surface and the catenoid are the only complete minimal surfaces of genus zero with total curvature -4π.

Hint. From Theorem 1 it follows that such a minimal surface has either one end of multiplicity three (simply-connected) or two ends of multiplicity one (doubly-connected). It remains to use Problem 5.

We now move directly to a description of the example of Meeks and Hoffman of a complete minimal surface of genus g embedded in $\mathbb{R}^3$. At the base of this construction lies Costa's example (Doctoral thesis, IMPA, Rio de Janiero, Brasil, 1982) of an immersion of a two-dimensional torus with three deleted points as a complete minimal surface with total curvature -12π, and therefore having three embedded ends. Moreover, all these ends are parallel, and one of them, say c_1, lies between the other two, c_2 and c_3: the end c_1 asymptotically tends to the plane $x^3 = \text{const}$, and c_2 and c_3 tend to $+\infty$ and $-\infty$ respectively in the direction of the x^3-axis, by analogy with the ends of the catenoid (see Figure 46).

Thus, all three embedded ends are disjoint, and we can hope that this minimal surface is embedded. In fact, Meeks and Hoffman, using the symmetry of the Costa surface, succeeded in proving that it is embedded. Later they generalized Costa's example and constructed complete embedded minimal surfaces of arbitrary genus with three ends and finite total curvature.

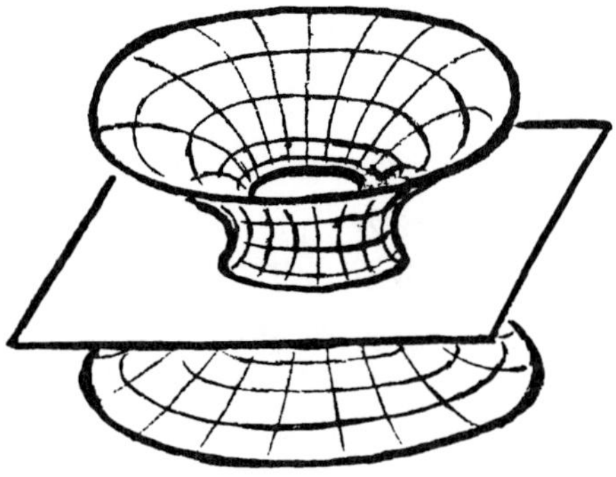

FIGURE 46

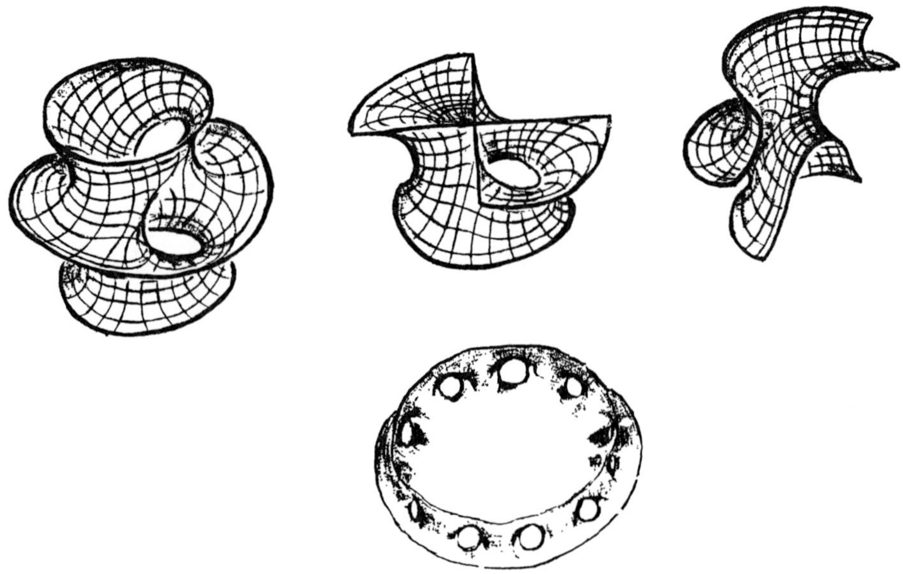

FIGURE 47

Figure 47 shows examples of surfaces of genus two and genus nine.

We mention that a computer played an important role in the research carried out by Meeks and Hoffman. The general idea of using computer graphics to investigate minimal surfaces for embeddedness is as follows. If it has been shown that all the ends of a complete minimal surface of finite total curvature are embedded, parallel, and tend asymptotically to different planes, then the investigation of embeddedness reduces to the investigation of a compact fragment of a minimal surface lying in a bounded domain (the size of which can be estimated). Using a computer, we can immediately see whether the surface under investigation has self-intersections in this finite domain or not. In the first case we can, again by using the computer, localize the self-intersections, that is, determine approximately where they lie and what they are. If no self-intersections are revealed, the computer can help, for example, to observe symmetries of the surface (if there are any). This is how Meeks and Hoffman used the computer. Thus, computer graphics enable us to make conjectures about the object under investigation, and this significantly lightens the problem.

We now touch briefly on the main ideas of the construction. For this we represent the torus T^2 as a unit square with opposite sides identified. It is convenient to regard the torus T^2 as the quotient of the complex plane c by the integral lattice $\mathbb{Z}^2$, $T^2 = \mathbb{C}/\mathbb{Z}^2$: two points z_1 and z_2 of $\mathbb{C}$ are identified if and only if $z_1 - z_2 = m + in$ for some integers m and n. In such a realization we can identify functions on T^2 with doubly-periodic (with respect to $\mathbb{Z}^2$) functions on $\mathbb{C}$, that is, with functions $f\colon \mathbb{C} \to \mathbb{C}$ such that for any integers m and n and any $z \in \mathbb{C}$ we have $f(z+m+in) = f(z)$. Henceforth we shall denote points of $\mathbb{Z}^2$ by the letter Ω, $\Omega = m + in$.

From the maximum principle it follows that any holomorphic function on a torus (and more generally on any closed Riemann surface) is a constant, since the modulus of such a function attains its maximum (in view of the compactness) at an internal point because of the absence of a boundary.

If f is a meromorphic function on T^2, it is not difficult to show that the sum of the residues of f over all the poles must be equal to zero. Therefore the simplest nonconstant meromorphic function on a torus has one pole of order two.

DEFINITION. We define the Weierstrass P-function as a meromorphic doubly-periodic function on $\mathbb{C}$ (with respect to $\mathbb{Z}^2$) having only double poles situated only at points of $\mathbb{Z}^2$ and given by

$$P(z) = \frac{1}{z^2} + \sum_{\substack{\Omega \neq 0 \\ \Omega = m+in}} \left[\frac{1}{(z-\Omega)^2} - \frac{1}{\Omega^2}\right].$$

It is easy to show that any two meromorphic doubly-periodic functions on $\mathbb{C}$ with respect to $\mathbb{Z}^2$ having only double poles at points of $\mathbb{Z}^2$ differ from one another by multiplication by a constant and addition of a (different) constant. In fact, some linear combination of such functions is a doubly-periodic function on $\mathbb{C}$ and everywhere holomorphic, so it is equal to a constant.

We have talked about the Weierstrass P-function becausec Costa used it in his construction. More concretely, let us cut out from the torus T^2 three points P_0, P_1, and P_2 corresponding to 0, $\omega_1 = 1/2$, and $\omega_2 = i/2$ respectively. Let $D = T^2 \setminus \{P_0, P_1, P_2\}$. We put $a = 2\sqrt{2\pi}P(1/2)$.

THEOREM (Costa; see [16]). *If P' denotes the derivative of the Weierstrass P-function P, and the domain $D \subset T^2$ and the constant a are defined as above, then the pair $(P\,dz, a/P')$ on D is a well-defined Weierstrass representation and gives a complete minimal surface with total curvature -12π.*

REMARK. The constant $a \in \mathbb{R}$ is the only constant $c \in \mathbb{C}$ for which the pair $(P dz, c/P')$ on D correctly defines a Weierstrass representation, that is, the corresponding holomorphic 1-forms φ^1, φ^2, and φ^3 have no real periods.

We shall not give the proof of this assertion here, or of the fact that the complete minimal surface constructed by Costa is actually embedded, as Meeks and Hoffman showed, using the fact that the Weierstrass P-function has a large number of symmetries. A reader who is interested can turn to [16].

§7. Indices of two-dimensional minimal surfaces in $\mathbb{R}^3$

In this section we touch on questions of stability of minimal surfaces and their bifurcations and the problem of calculating the indices—an important characteristic of minimal surfaces, closely connected with the first two questions. As a rule, we shall not give proofs, since the admitted complexity of our book and limitations of space do not allow us to do this.

In §1 of Chapter 2 we described in detail the reconstruction (bifurcation) of catenoids as the distance between the bounding circles varies. How can we characterize the positions of a contour for which bifurcations occur? Why does only one catenoid occur in practice?

In §2 of Chapter 1 we mentioned that a soap film always tends to take up a form having (locally) minimal area. Now suppose that we have succeeded in realizing as a soap film a certain minimal surface for which there is a deformation that monotonically decreases the area of this surface. In this case small fluctuations of the film, which always exist in the real world, lead to spasmodic modification of its form—the soap film will be unstable. This observation can serve as motivation for the term "unstable minimal surface" for such surfaces. In practice it is very difficult to obtain unstable films (see [2]).

Next, suppose we have a soap film M spanning a wire contour Γ. If we deform Γ, the film M is also deformed. As we have already observed in the example of the catenoid, not every continuous change of form of a contour leads to a continuous change of form of the soap film spanning it. The reason for the spasmodic modifications that occur here is that at the time of deformation of the contour Γ the film that is deformed with it may become unstable. An unstable film starts to be modified and continues until it becomes stable. Generally speaking, its topological type and connectivity may change (as, for example, in the case of the catenoid). In this connection we naturally wish to investigate whether the given minimal surface is stable, and if it is unstable to determine the degree of instability.

For a mathematical study of stability of a minimal surface we proceed, as we have already mentioned in §2 of Chapter 1, by analogy with the investigation of the extrema of smooth functions. For this we usually calculate the second derivative of the area functional—the Hessian, which gives a bilinear form on the space of admissible small variations of the surface—and see whether the Hessian is positive definite or not. If not, then as in the case of functions we calculate the null-index and the index, which characterize the degree of degeneracy and negative definiteness of the Hessian.

The indices of different variational functionals have been, particularly recently, the object of intensive research. There is most interest in the area functional (in the multidimensional case the volume functional) and the Dirichlet (energy) functional (see [17], [18], [19], [4]).

Let us give some results that touch on the investigation of stability of two-dimensional minimal surfaces in $\mathbb{R}^3$. Barbosa and Do Carmo showed (see [20]) that an orientable minimal surface in $\mathbb{R}^3$ for which the area of the image under the Gaussian mapping is less than 2π is stable. In [21] Do Carmo and Peng gave a generalization of Bernstein's problem in terms of stability: it turns out that the plane is also the only complete stable minimal surface in $\mathbb{R}^3$ (we recall that the plane is the only minimal surface that is a complete graph, that is, the graph of a function defined on the whole plane; see §3 of Chapter 2).

In the case of a compact minimal surface the index is always finite (see below), but for noncompact minimal surfaces this is not always so. In the papers mentioned below the authors determine when the index of a complete minimal surface is finite. We recall that the total curvature τ of a two-dimensional orientable surface $\psi\colon M \to \mathbb{R}^3$ is defined as the integral over the surface of its Gaussian curvature K: $\tau = \int_M K$. In [22] Fischer-Colbrie proved that for complete immersed orientable minimal surfaces in $\mathbb{R}^3$ the finiteness of the index is equivalent to the finiteness of the total curvature. In [23] Tysk gave an estimate of the index $\operatorname{ind}(M)$ of such a complete minimal surface $\psi\colon M \to \mathbb{R}^3$ in terms of the Gaussian mapping. It turns out that $\operatorname{ind}(M) \leq 7.68183 \cdot k$, where k is the degree of the Gaussian mapping.

The calculation of the indices of noncompact minimal surfaces is a complicated problem, and until recently the investigation of them has been restricted by the determination of the stability, that is, determining whether the index is zero or not. The first numerical values of indices were apparently obtained in 1985 by Fischer-Colbrie (see [22]), where, apart from anything else, she showed that the indices of two classical minimal surfaces, namely the catenoid and the Enneper surface, are equal to one.

In [24] Lopez and Ros, relying on Fischer-Colbrie's results, showed that the catenoid and the Enneper surface are the only complete orientable minimal surfaces in $\mathbb{R}^3$ with index one. This result is one more characteristic of these two classical minimal surfaces: as Osserman showed (see [9]), the catenoid and the Enneper surface are the only complete minimal surfaces of total curvature -4π; they are also the only complete minimal surfaces for which the Gaussian mapping is a diffeomorphism with an image.

In [25] Rassias made an attempt to calculate the indices of the classical minimal surfaces in $\mathbb{R}^3$, but it was not crowned with success, because the author confused the Jacobi equation for a minimal surface written in an isothermal coordinate system (where the Euclidean Laplacian is used) with the Jacobi equation "on the surface" (where the metric Laplacian is used).

In this section we consider the problem of calculating the index by an example of the area functional of two-dimensional minimal surfaces in $\mathbb{R}^3$.

Let us consider the definition of the index in more detail. We first define the index of a compact minimal surface, and then generalize this definition to the noncompact case. Thus, to start with let $\psi\colon M \to \mathbb{R}^3$ be an arbitrary immersed surface with smooth boundary $\psi\colon \partial M \to \mathbb{R}^3$. We define a *variation* of this surface as a smooth one-parameter family of immersions $\psi_t\colon M \to \mathbb{R}^3$ such that $\psi_0 = \psi$, $0 \le t \le 1$, and the field of the variation is the mapping $\xi\colon M \to \mathbb{R}^3$ that assigns to each point $P \in M$ the vector $\xi(P)$ of $\mathbb{R}^3$ that is equal to the velocity of the motion of the point $\psi_t(P)$ under the action of the variation ψ_t at the initial instant:

$$\xi(P) = d\psi_t(P)/dt|_{t=0}.$$

Suppose that the manifold M is orientable and compact. Then we can calculate the area $A(\psi(M))$ of M in the metric ds^2 induced by the immersion ψ (the area of the surface $\psi\colon M \to \mathbb{R}^3$). An arbitrary variation $\psi_t\colon M \to \mathbb{R}^3$ of this surface induces on M a one-parameter family of metrics ds_t^2, $ds_{t=0}^2 = ds^2$. The area $A(t)$ of M in the metric ds_t^2 (the area of the surface $\psi_t\colon M \to \mathbb{R}^3$) is a smooth function of the parameter t. To justify the equivalence of the two definitions of a minimal surface as a surface of zero mean curvature and as a critical point of the area functional it is sufficient to calculate the derivative $dA(t)/dt|_{t=0}$.

Let n be the field of unit normals to the surface $\psi\colon M \to \mathbb{R}^3$, that is, the mapping $n\colon M \to \mathbb{R}^3$ that assigns to each point $P \in M$ one (chosen smoothly) of the two unit vectors of $\mathbb{R}^3$ orthogonal to the image $\psi_*(T_PM)$ of the tangent plane T_PM. Let H denote the mean curvature of this surface with respect to n.

PROPOSITION 1. *If $\psi\colon M \to \mathbb{R}^3$ is an arbitrary immersed compact orientable surface, and $\psi_t\colon M \to \mathbb{R}^3$ is a variation of it that is fixed on the boundary ∂M of M (∂M may be empty), and $A(t)$ is the area of the surface $\psi_t\colon M \to \mathbb{R}^3$ for each t, then*

$$dA(t)/dt|_{t=0} = -\int_M 2H\langle \xi, n\rangle = \int_M 2HT,$$

where ξ is the field of the variation ψ_t, $T \cdot n$ is its normal component, $T = \langle \xi, n\rangle$, and the surface integral of a function is calculated by means of the area form in the metric ds^2 induced by the immersion $\psi_0 = \psi\colon M \to \mathbb{R}^3$.

REMARK. The condition that the variation ψ_t is fixed on the boundary ∂M imposes a restriction on the function T: $M \to \mathbb{R}^3$: $T|_{\partial M} \equiv 0$.

Now it is not difficult to show that the vanishing of the derivative $dA(t)/dt|_{t=0}$ for any variation fixed on the boundary is equivalent to the vanishing of the mean curvature H (compare with §2 of Chapter 3).

Let ψ: $M \to \mathbb{R}^3$ be an orientable compact immersed minimal surface. Consider a two-parameter variation $\psi_{ts} : M \to \mathbb{R}^3$ of our surface, $\psi_{ts}|_{t=0,s=0} = \psi$, that is, a smooth mapping F: $I \times I \times M \to \mathbb{R}^3$, $\psi_{ts}(P) = F(t, s, P)$ which for each fixed t and s is an immersion and $F(0, 0, P) = \psi(P)$. Let us construct the function $A(t, s)$ equal to the area of the surface $\psi_{ts} : M \to \mathbb{R}^3$, that is, the area of M in the metric ds^2_{ts} induced by the immersion $\psi_{ts} : M \to \mathbb{R}^3$. Our problem is to calculate the second mixed derivative of $A(t, s)$ at the initial moment of deformation, that is, $\partial^2 A(t, s)/\partial t \partial s|_{t=0,s=0}$.

Let ξ denote the field of the variation ψ_{t0}, and η the field of the variation ψ_{0s}, that is,

$$\xi(P) = \left.\frac{\partial \psi}{\partial t}\right|_{t=0,s=0}, \qquad \eta(P) = \left.\frac{\partial \psi}{\partial s}\right|_{t=0,s=0}.$$

PROPOSITION 2. *If ψ: $M \to \mathbb{R}^3$ is an arbitrary immersed compact orientable minimal surface, ψ_{ts}: $M \to \mathbb{R}^3$ a two-parameter variation of it that is fixed on the boundary ∂M, and $A(t, s)$ the area of the surface ψ_{ts}: $M \to \mathbb{R}^3$ for each t and s, then*

$$\left.\frac{\partial^2 A(t, s)}{\partial t \partial s}\right|_{t=0,s=0} = -\int_M (\Delta T - 2KT) \cdot S,$$

where $T = \langle \xi, n \rangle$, $S = \langle \eta, n \rangle$, and n is the field of unit normals to the surface ψ: $M \to \mathbb{R}^3$. Here Δ denotes the so-called metric Laplacian. If (u, v) are isothermal coordinates, $ds^2 = \lambda(du^2 + dv^2)$, then $\Delta T = (1/\lambda) \cdot (\partial^2 T/\partial u^2 + \partial^2 T/\partial v^2)$.

REMARK 1. It is easy to see that the standard Euclidean Laplacian $\partial^2/\partial u^2 + \partial^2/\partial v^2$, written in the (u, v)-coordinates, is changed by a change of coordinates, that is, in the new coordinates (u_1, v_1) this operator is not equal to $\partial^2/\partial u_1^2 + \partial^2/\partial v_1^2$ (if we require that its value on each smooth funciton, written in the new coordinate system, is unchanged).

The metric Laplacian Δ is an invariant differential operator acting on functions (independent of the coordinate system), and if in the (u, v)-coordinates the metric $g_{ij} = \delta_{ij}$, then this operator coincides with the Euclidean Laplacian $\partial^2/\partial u^2 + \partial^2/\partial v^2$. In the general form the metric Laplacian is

$$\Delta T = (1/\sqrt{g}) \cdot \partial/\partial u^i(\sqrt{g} g^{ij} \partial T/\partial u^j),$$

where (u^1, u^2) are local coordinates, $T(u^1, u^2)$ is a smooth function, (g^{ij}) is the matrix inverse to the matrix of the metric (g_{ij}), and $g = \det(g_{ij})$.

Here we always assume summation over repeated indices. If (u, v) are isothermal coordinates, $u = u^1$, $v = u^2$, and $g_{ij} = \lambda\delta_{ij}$, then $\Delta T = (1/\lambda)\cdot(\partial^2 T/\partial u^2 + \partial^2 T/\partial v^2)$ (verify this).

REMARK 2. From Propositions 1 and 2 it follows that the first and second differentials of the area functional depend only on the normal components of the fields of the variations, and not on the tangential components. This is easily demonstrated by the following example.

Let ξ be the field of some variation of an immersed surface $\psi\colon M \to \mathbb{R}^3$, fixed on the boundary. Suppose that the field ξ touches this surface, that is, for any point P the vector $\xi(P)$ of $\mathbb{R}^3$ lies in the image $\psi_*(T_P M)$ of the tangent plane $T_P M$. Since the tangent mapping $\psi_*\colon T_P M \to \mathbb{R}^3$ is an embedding for any point $P \in M$, we can construct a smooth field $\eta = \psi_*^{-1}(\xi)$ on M, where $\eta = 0$ on ∂M. The field η generates a local one-parameter group of diffeomorphisms $\varphi_t\colon M \to M$ such that φ_0 is the identity transformation, and for any point $P \in M$ the tangent vector to the curve $\varphi_t(P)$ is equal to η when $t = 0$. The curves $\gamma(t) = \varphi_t(P)$ are called the *integral curves* of this field (see [1]).

If we now consider the variation $\psi = \psi \cdot \varphi_t\colon M \to \mathbb{R}^3$ which, simply speaking, is a change of parametrization of the surface $\psi\colon M \to \mathbb{R}^3$, then the field of this variation is precisely equal to η.

Clearly, a change of parametrization does not change the area of the surface $\psi\colon M \to \mathbb{R}^3$, so $A(t) = A(\psi_t(M)) = \text{const}$, and any derivative of $A(t)$ with respect to t is equal to zero.

From now on we shall only be interested in normal variations, that is, variations for which the velocity field at the initial instant is orthogonal to the surface. If M is orientable and n is the field of unit normals to the immersed surface $\psi\colon M \to \mathbb{R}^3$, then the field ξ of any normal variation fixed on the boundary is given by means of the function $T\colon M \to \mathbb{R}^3$, $\xi(P) = T(P)\cdot n(P)$, $T|_{\partial M} \equiv 0$. We denote the linear space of such functions by $C_0^\infty(M)$.

Thus, the Hessian

$$I(T, S) = -\int_M (\Delta T - 2KT)\cdot S = \int_M J(T)\cdot S$$

in Proposition 2 is a bilinear form on the space $C_0^\infty(M)$. Here we have denoted by J the differential operator $-\Delta + 2K$, $J\colon C_0^\infty(M) \to C^\infty(M)$ called the *Jacobi operator*, where $C^\infty(M)$ is the space of smooth functions on M.

On the space $C_0^\infty(M)$ we can introduce the standard scalar product defined as $\langle T, S\rangle = \int_M T\cdot S$.

PROPOSITION 3 (see [26]). *The Hessian I is a symmetric bilinear form on the space $C_0^\infty(M)$. The form I can be expressed by means of a diagonal matrix in terms of the standard scalar product and has eigenvalues $\{\lambda_i\}$ such* *that*

$$\lambda_1 < \lambda_2 < \lambda_3 < \cdots \to +\infty.$$

In addition, to each real eigenvalue λ_i *there corresponds a finite-dimensional eigensubspace* $V(\lambda_i)$.

DEFINITION. The *index* $\operatorname{ind}(M)$ of an immersed compact orientable minimal surface $\psi\colon M \to \mathbb{R}^3$ is defined as the index of the Hessian $I(T, S)$ of the area functional as a bilinear form, that is, the sum of the dimensions of the eigensubspaces $V(\lambda_i)$ corresponding to the negative eigenvalues: $\operatorname{ind}(M) = \sum_{\lambda_i<0} \dim V(\lambda_i)$.

The *null-index* of a minimal surface is defined as the null-index of the Hessian I, that is, the dimension of the eigensubspace $V(0)$ corresponding to the zero eigenvalue (the dimension of the kernel of the bilinear form I).

If the null-index is not equal to zero, then we call the boundary ∂M *conjugate*, and the null-index is the *multiplicity* of this conjugate boundary. We call the Hessian itself the *index form.*

The condition that $T \in C_0^\infty(M)$ lies in the kernel of the index form is equivalent to T being a solution of the equation $\Delta T - 2KT = 0$, which is called the Jacobi equation. The fields of variations $T \cdot n$ corresponding to solutions T of this equation are called *Jacobi fields.* Clearly, all Jacobi fields that vanish on ∂M form a linear space whose dimension is either zero or equal to the multiplicity of the conjugate boundary.

The index and the null-index characterize the degree of instability of a minimal surface. If they are both zero, then the minimal surface is stable: any nonzero normal variation fixed on the boundary increases the area of this surface.

Above we gave a definition of the index of a compact minimal surface. However, the basic classical examples of two-dimensional minimal surfaces in $\mathbb{R}^3$ are complete, and consequently noncompact (see above). For noncompact minimal surfaces we can also introduce the concept of the index; we now turn to a definition of it. We first observe that any subdomain of a minimal surface is also a minimal surface.

DEFINITION. The *index* $\operatorname{ind}(M)$ of a noncompact orientable immersed minimal surface $\psi\colon M \to \mathbb{R}^3$ is defined as the least upper bound of the indices of minimal surfaces of the form $\psi\colon K \to \mathbb{R}^3$, where $K \subset M$ is an arbitrary compact subdomain of M having a smooth boundary (by a compact domain we mean here a domain with compact closure).

Note that a noncompact minimal surface is said to be stable if its index is zero.

The motivation for the definition of the index just given is a theorem of Smale and Simons, which describes the behavior of the eigenvalues of the restriction I_k of the index form I to the space $C_0^\infty(K)$ under a contraction of K. This theorem enables us to reduce the calculation of the index of a minimal surface to the solution of the Jacobi equation. First of all we give a definition of a contraction.

DEFINITION. Let K be a compact subdomain of M with smooth boundary. We define a *contraction* of K as a family $\{K_t\}$ of subdomains of M with smooth boundaries, $t \in [a, b]$, such that $K_a = K$ and if $t > s$, then $K_t \subsetneqq K_s$. A contraction is said to be *smooth* if the boundary ∂K_t of the domain K_t depends smoothly on t.

THEOREM (Smale, Simons; see [27], [26]). *Let $\psi\colon M \to \mathbb{R}^3$ be a compact orientable immersed minimal surface with smooth boundary and $\{M_t\}$, $t \in [a, b]$, a smooth contraction. Then the eigenvalues $\lambda_1(t) \le \lambda_2(t) \le \cdots \to +\infty$ of the index form I_t, the restriction of the index form I to $C_0^\infty(M_t)$, are continuous strictly monotonically increasing functions of the parameter t (here the number of times each eigenvalue occurs is equal to the dimension of the corresponding eigensubspace).*

Moreover, there is an $\varepsilon > 0$ such that if K is an arbitrary compact subdomain of M having a smooth boundary, and the area $A(\psi(K))$ of the minimal surface $\psi\colon K \to \mathbb{R}^3$ is less than ε, $A(\psi(K)) < \varepsilon$, then all the eigenvalues of the index form I_k are strictly positive. If the contraction is not smooth, then the eigenvalues are again strictly monotonically increasing, but not necessarily continuous.

DEFINITION. We call a contraction $\{M_t\}$ a *contraction of ε-type* if for sufficiently large t the area of the surface $\psi\colon M \to \mathbb{R}^3$ is less than ε.

COROLLARY 1 (Smale, Simons; see [27], [26]). *Let $\psi\colon M \to \mathbb{R}^3$ be a smooth compact orientable immersed minimal surface with smooth boundary. Then*

a) *for any contraction $\{M_t\}$, $t \in [a, b]$, of the manifold M (the closure of the initial domain M_a coincides with M) only finitely many minimal surfaces $\psi\colon M_t \to \mathbb{R}^3$ have conjugate boundaries ∂M_t;*

b) *there is an $\varepsilon > 0$ such that for any smooth contraction of ε-type $\{M_t\}$, $t \in [a, b]$, of the manifold M the index of the surface $\psi\colon M \to \mathbb{R}^3$ is equal to the sum of the multiplicities of the conjugate boundaries ∂M_t over all $t \in (a, b]$. More precisely, if $\beta(t) = \dim\ker I_t$, where I_t is the restriction of the index form I to the space $C_0^\infty(M_t)$, then $\operatorname{ind}(M) = \sum_{a<t\le b} \beta(t)$.*

If the contraction is not smooth, then $\operatorname{ind}(M) \ge \sum_{a<t\le b} \beta(t)$.

We note that from Corollary 1 it follows that the index of any compact minimal surface is finite.

In the noncompact case the role of a smooth contraction of ε-type is played by a smooth exhaustion.

DEFINITION. An *exhaustion* of an orientable immersed noncompact minimal surface $\psi\colon M \to \mathbb{R}^3$ is defined as a family $\{M_t\}$, $t \in (0, \infty)$, of compact subdomains of M having smooth boundaries such that

a) $M_t \subsetneqq M_s$ so long as $t < s$;

b) for any compact subset $K \subset M$ there is a t such that $K \subset M_t$;

c) the indices and the null-indices of the surfaces ψ: $M_t \to \mathbb{R}^3$ are zero for sufficiently small t.

We call an exhaustion *smooth* if the boundary ∂M_t of the domain M_t depends smoothly on t.

COROLLARY 2 (see [29]). *Let* ψ: $M \to \mathbb{R}^3$ *be a smooth noncompact orientable immersed minimal surface, and* $\{M_t\}$, $t \in (0, \infty)$, *a smooth exhaustion of it. Then the index of the surface* ψ: $M \to \mathbb{R}^3$ *is equal to the sum of the multiplicities of the conjugate boundaries* ∂M_t *over all* $t \in (0, \infty)$. *More precisely, if* $\beta(t) = \dim \ker I_t$, *where* I_t *is the restriction of the index form* I *to the space* $C_0^\infty(M_t)$, *then* $\operatorname{ind}(M) = \sum_{0<t<\infty} \beta(t)$.

If the exhaustion is not smooth, then $\operatorname{ind}(M) \geq \sum_{0<t<\infty} \beta(t)$.

Thus, to calculate the index of a compact (noncompact) minimal surface ψ: $M \to \mathbb{R}^3$ it is sufficient to construct a smooth contraction of ε-type (respectively, a smooth exhaustion) $\{M_t\}$, to find the solution of the Jacobi equation for each M_t, and from all the solutions to choose only those that vanish on ∂M_t. This program enables us to calculate the indices of a whole series of classical minimal surfaces. Before stating the main result of this section, we give one more assertion that is very useful in the calculation of the indices of two-dimensional minimal surfaces in $\mathbb{R}^3$.

We first observe the following. Let $(f(w), g(w))$ over $U \subset \mathbb{C}$ be the local Weierstrass representation of an immersed orientable minimal surface ψ: $M \to \mathbb{R}^3$, $w = u + iv$; then (u, v) are isothermal coordinates, and the induced metric ds^2 has the form $ds^2 = \lambda(du^2 + dv^2)$. From Proposition 2 it follows that the Jacobi equation in the (u, v)-coordinates has the form

$$\partial^2 T/\partial u^2 + \partial^2 T/\partial v^2 - 2K\lambda T = 0,$$

where K is the Gaussian curvature. In §3 we establish that

$$\lambda = |f|^2(1 + |g|^2)^2,$$

and $K = -4|dg/dw|^2/|f|^2(1+|g|^2)^4$, so in terms of the functions f and g that specify the Weierstrass representation the Jacobi equation is written as

$$\partial^2 T/\partial u^2 + \partial^2 T/\partial v^2 + [8|dg/dw|^2/(1 + |g|^2)^2] \cdot T = 0.$$

Thus, the Jacobi equation of a minimal surface in $\mathbb{R}^3$ does not depend on f, so it depends only on the Gaussian mapping of this surface. This remark enables us to state the following proposition (Tuzhilin).

PROPOSITION 4 (see [29]). *Let* ψ_k: $M \to \mathbb{R}^3$, $k = 1, 2$, *be two immersed orientable minimal surfaces. Suppose there is a diffeomorphism* F: $M \to M$ *compatible with the Gaussian mappings of the two surfaces. This means that if* n_k *is the Gaussian mapping of the minimal surface* ψ_k: $M \to \mathbb{R}^3$, *then* $n_2 \circ F = \varphi \circ n_1$, *where* φ: $S^2 \to S^2$ *is an isometry of the sphere* S^2. *It is convenient to specify this condition by means of the commutative diagram*

$$\begin{array}{ccc} M & \xrightarrow{n_1} & S^2 \\ {\scriptstyle F}\big\downarrow & & \big\downarrow{\scriptstyle \varphi} \\ M & \xrightarrow{n_2} & S^2 \end{array}$$

Then these surfaces have identical indices.

In all five theorems of Tuzhilin given below, S denotes a noncompact immersed orientable two-dimensional connected minimal surface $\psi\colon M \to \mathbb{R}^3$.

The first theorem gives a necessary and sufficient condition for the finiteness of the index of a minimal surface S given by the Weierstrass representation (M, ω, g) for some class of functions g called good (for the definition, see below). This theorem follows from the results of Fischer-Colbrie stated above. In Corollary 1 we give a criterion for the finiteness of the index of a minimal surface $\psi\colon \mathbb{C} \to \mathbb{R}^3$ given by the Weierstrass representation $(\mathbb{C}, f\,dw, g)$, where g is a good function. In Corollary 2 we give the most general form of good functions known to the authors of this book.

DEFINITION. We say that a meromorphic function g, defined on a Riemann surface M, is *good* if there is a 1-form ω holomorphic on M such that the Weierstrass representation (M, ω, g) defines a complete immersed minimal surface.

THEOREM 1. *Suppose that a minimal surface S is given by the Weierstrass representation (M, ω, g), where g is a good function. Then the index of S is finite if and only if the total curvature of S is finite.*

COROLLARY 1. *Suppose that a minimal surface S is given by the Weierstrass representation $(\mathbb{C}, f\,dw, g)$, where g is a good function. Then the index of S is finite if and only if g is a rational function.*

COROLLARY 2. *Suppose that a minimal surface S is given by the Weierstrass representation $(\mathbb{C}, f\,dw, g)$, and that g is either of the form $g = P/h + c/Q$ or of the form $g = h/P$, where P and Q are arbitrary polynomials, $c \in \mathbb{C}$, and h is an arbitrary holomorphic function. Then the index of S is finite if and only if g is a rational function.*

REMARK. The class of good functions described in Corollary 2 is quite wide. It contains all meromorphic functions having finitely many zeros or finitely many poles; all rational functions; all trigonometric and hyperbolic functions, for example,

$$\tan(z) = \sin(z)/\cos(z) = (e^{iz} - e^{-iz})/(e^{iz} + e^{-iz}) = 1 - 1/e^{iz}\cos(z),$$
$$\tan(z) = (e^{z} - e^{-z})/(e^{z} + e^{-z}) = 1 - 1/e^{z}\cosh(z);$$

all meromorphic functions that take a certain value with finite multiplicity or do not take a certain value at all (in fact, if c is such a value, then the function $g - c$ can be represented in the form F/h, so $g = P/h + c)$. As an illustration of the last assertion we consider a partition of the function e^z into a sum of two holomorphic functions φ and ψ: $e^z = \varphi + \psi$. Since $e^z \neq 0$, the meromorphic function $g = \varphi/\psi$ does not take the value -1, so $g + 1 = 1/h$, where $h = 1/(g+1)$ is holomorphic, and therefore g is equal to $1/h - 1$ and is a good function.

THEOREM 2. *Suppose that a minimal surface S is given by one of the following Weierstrass representation:*

a) $(\mathbb{C}, \omega, (aw+b)^m)$, $a, b \in \mathbb{C}$, $a \neq 0$, *where m is a natural number;*

b) $(\mathbb{C}\backslash\{-b/a\}, \omega, (aw+b)^m)$, $a, b \in \mathbb{C}$, $a \neq 0$, *where m is a nonzero integer.*

Then the index of S is equal to $2|m| - 1$.

THEOREM 3. *Suppose that a minimal surface S is given by the Weierstrass representation $(U, \omega, (aw+b)^m)$, $a, b \in \mathbb{C}$, $a \neq 0$, where m is a nonzero integer and $U \subset \mathbb{C}$ is a subdomain of the complex plane $\mathbb{C}$ (if m is negative, we assume that the point $\{-b/a\}$ does not lie in the domain U). Then the index of S does not exceed* $2|m| - 1$.

Next we define a closed subset K of the standard Euclidean sphere $S^2 \subset \mathbb{R}^3$ with center at the origin in one of the following ways:

a) when $m > 0$ we put $K = S^2 \cap \{x^3 \leq (m-1)/m\}$; if $m = 1$, we take for K a closed hemisphere not containing the North Pole;

b) when $m \in \mathbb{Z}\backslash\{0\}$ let K be the part of the sphere S^2 not containing the North Pole and enclosed between two parallel noncoincident planes situated at a distance $\tanh(t_{m-1})$ from the center of the sphere. Here t_{m-1} denotes the only positive root of the equation $(m-1)/m = \tanh(t)\cdot\tanh(t\cdot(m-1)/m)$.

THEOREM 4. *Suppose that a minimal surface S is given by the Weierstrass representation $(U, \omega, (aw+b)^m)$, $a, b \in \mathbb{C}$, $a \neq 0$, where m is a nonzero integer and $U \subset \mathbb{C}$ is a subdomain of the complex plane $\mathbb{C}$ (if m is negative, we assume that the point $\{-b/a\}$ does not lie in U). Suppose that U contains the inverse image under the Gaussian mapping of the subset K of S^2 defined in either* a) *or* b) *above. Then the index of S is equal to* $2|m| - 1$.

Let us define an open subset K' of the standard unit sphere $S^2 \subset \mathbb{R}^3$ with center at the origin in one of the following ways:

a$'$) $K' = S^2 \cap \{\overline{\overline{X}}^3 \leq 0\}$; if $m = 1$, we take for K' an open hemisphere not containing the North Pole;

b$'$) K' is the part of S^2 not containing the North Pole and enclosed between two parallel noncoincident planes situated at a distance $\tanh(t_0)$ from the center of the sphere. Here t_0 denotes the only positive root of the equation $t \cdot \tanh(t) = 1$.

THEOREM 5. *Suppose that a minimal surface S is given by the Weierstrass representation $(U, \omega, (aw+b)^m)$, $a, b \in \mathbb{C}$, $a \neq 0$, where m is a nonzero* *integer and $U \subset \mathbb{C}$ is a subdomain of the complex plane $\mathbb{C}$ (if m is negative, we assume that the point $\{-b/a\}$ does not lie in U). Suppose that the image of U under the Gaussian mapping is contained in the subset K' of the sphere S^2 defined above in* a′) *or* b′). *Then the index of S is equal to zero.*

We give the numerical values of the indices of the classical minimal surfaces, obtained by means of Theorems 1–5 (A. A. Tuzhilin).

COROLLARY.

Surface	*Weierstrass representation*	*Index*
Enneper surface	$(\mathbb{C}, dw, w)$	1
Catenoid	$(\mathbb{C}\backslash\{0\}, dw/(2w^2), w)$	1
Richmond surface	$(\mathbb{C}\backslash\{0\}, w^2 dw, 1/w^2)$	3
Incomplete Scherk surface	$(U = \{\lvert w\rvert < 1\}, dw/(1-w^4), w)$	0
Helicoid	$(\mathbb{C}, -ie^{-\omega}dw, e^{\omega})$	∞

The indices of all periodic minimal surfaces, in particular, the Schwarz-Riemann surface, and also the complete Scherk surface, are equal to infinity.

In §1 of Chapter 2 we promised to prove that a bifurcation of the catenoid occurs on the conjugate boundary, that is, when the catenoid loses its stability. We now fulfill our promise, at the same time illustrating the method of proof of Theorem 2 by the example of the catenoid. We also show that of the two catenoids stretched on wire circles (see §1 of Chapter 2) one is stable and the other unstable.

Let (r, φ, z) be cylindrical coordinates in $\mathbb{R}^3$ in which the catenoid $\psi: M \to \mathbb{R}^3$ is written as $r = \cosh(z)$. Let (φ, z) be coordinates on the catenoid; then it is easy to calculate that the Jacobi equation has the form

$$\partial^2 T/\partial z^2 + \partial^2 T/\partial \varphi^2 + [2/\cosh^2(z)] \cdot T = 0.$$

Consider a smooth exhaustion of the catenoid by domains $M_t = \{\lvert z\rvert < t\}$. For each z we expand $T(\varphi, z)$ as a Fourier series in φ:

$$T(\varphi, z) = a_0(z)/2 + \sum a_k(z)\cos(k\varphi) + b_k(z)\sin(k\varphi).$$

Since T is assumed to be smooth with respect to φ and z, the coefficients $a_0(z)$, $a_k(z)$, and $b_k(z)$ are also smooth functions of z. Substituting T into the Jacobi equation, we obtain a system of ordinary differential equations for the coefficients:

$$\begin{aligned} &a_0'' + [2/\cosh^2(z)] \cdot a_0 = 0, \\ &a_k'' + [2/\cosh^2(z) - k^2] \cdot a_k = 0, \qquad k \geq 1, \\ &b_k'' + [2/\cosh^2(z) - k^2] \cdot b_k = 0, \qquad k \geq 1. \end{aligned}$$

The solution of these equations can be written explicitly:

$$a_0 = c_1(z \cdot \tanh(z) - 1) + c_2 \tanh(z),$$
$$a_1 = c_1(1/\cosh(z)) + c_2(\sinh(z) + z/\cosh(z)),$$

when $k > 1$

$$a_k = c_1(\tanh(z)\sinh(kz) - k \cdot \cosh(kz)) + c_2(\tanh(z)\cosh(kz) - k \cdot \sinh(kz)),$$

(similarly for b_k).

Here c_1 and c_2 are arbitrary real constants.

From these equations we must choose a_k and b_k so that the function $T(\varphi, z) = a_0(z)/2+\sum a_k(z)\cos(k\varphi)+b_k(z)\sin(k\varphi)$, vanishes on the boundary of some M_t. This is equivalent to the vanishing of the coefficients a_k and b_k when $z = \pm z_0$ for some $z_0 > 0$.

We observe that each coefficient can be represented as a linear combination of an even and an odd function, and these two functions do not vanish simultaneously. Hence it follows that only the even component occurs in the solution. Thus, the second coefficient c_2 is zero.

When $c_1 \neq 0$ the function $a_1 = c_1/\cosh(z)$ does not vanish anywhere. When $k > 1$ the function $a_k = c_1(\tanh(z)\sinh(kz) - k \cdot \cosh(kz))$ also does not vanish anywhere, since $\tanh(z)\tanh(kz) < 1 < k$ (similarly for b_k, $k \geq 1$). Therefore the only candidate for the Jacobi field for our exhaustion is $a_0 = c_1(z \cdot \tanh(z) - 1)$.

It is easy to see that when $z > 0$ the function $z \cdot \tanh(z) - 1$ has a unique root, since when $z > 0$ there is a unique point of intersection of the graphs $y = \tanh(z)$ and $y = 1/z$.

Thus, the index of the catenoid $r = \cosh(z)$ is equal to one, and the one-dimensional space of Jacobi fields on M_{z_0}, where $z_0 \cdot \tanh(z_0) = 1$, $z_0 > 0$, is generated by the field $(z \cdot \tanh(z) - 1) \cdot n$, where n is the field of unit normals to the catenoid.

We now consider an arbitrary catenoid $r = a \cdot \cosh(z/a)$ with coordinates (φ, z). It is easy to calculate that the unit normal to such a catenoid is equal to $[1/\sinh(z/a)] \cdot (\cos(\varphi), \sin(\varphi), -\cosh(z/a))$, so the diffeomorphism F of the catenoids $r = \cosh(z)$ and $r = a \cdot \cosh(z/a)$ given by $F\colon (\varphi, z) \to (\varphi, az)$ is compatible with the Gaussian mappings. Therefore, by Proposition 4, the index of the catenoid $r = a \cdot \cosh(z/a)$ is also equal to one. Moreover, under this diffeomorphism the Jacobi field defined by the function $T = z \cdot \tanh(z) - 1$ over M_{z_0} goes into the Jacobi field defined by the function $T = (z/a) \cdot \tanh(z/a) - 1$ over M_{az_0}. Therefore for the exhaustion $M_t = \{|z| < t\}$ of the catenoid $r = a \cdot \cosh(z/a)$ only M_{az_0} has a conjugate boundary, where z_0 is the only positive root of the equation $z \cdot \tanh(z) = 1$.

REMARK. To calculate the index of the more general minimal surfaces described in part b) of Theorem 2, it is sufficient to consider, instead of the catenoid an m-sheeted covering of it, that is, an immersion ψ: $\mathbb{R} \times S^1 \to \mathbb{R}^3$ of the cylinder $\mathbb{R} \times S^1$ with coordinates (z, φ), $z \in \mathbb{R}$, $0 \leq \varphi < 2\pi$, defined by

$$x^1 = \cosh(z)\cos(k\varphi), \quad x^2 = \cosh(z)\cos(k\varphi), \quad x^3 = z,$$

where x^1, x^2, x^3 are the standard coordinates in $\mathbb{R}^3$.

LEMMA (Tuzhilin). *The index of the m-sheeted covering of the catenoid described above is equal to* $2m - 1$.

It remains to observe that between the minimal surfaces described by the Weierstrass representation in part b) of Theorem 2 and the m-sheeted covering of the catenoid there is a diffeomorphism compatible with the Gaussian mappings, and to apply Proposition 4.

In §1 of Chapter 2 we established that the parameters a of the catenoids $r = a \cdot \cosh(z/a)$ spanning two circles of radius ρ at a distance h apart are obtained as the coordinates (with respect to a) of the points of intersection of the graphs $y = a \cdot \cosh(h/2a)$ and $y = \rho$. The graph $y = a \cdot \cosh(h/2a)$ has a unique critical point $a_0 = h/2z_0$, where z_0 is the only positive root of the equation $\coth(z) = z$, or equivalently of the equation $z \cdot \tanh(z) = 1$.

Therefore, if $h = h_{\text{cr}}(\rho)$ is the distance between the circles for which the two catenoids stick together, then in this case the line $y = \rho$ touches the graph $y = a \cdot \cosh(h/2a)$ at the point (a_0, ρ) and the only catenoid stretched on this contour has the equation $r = a_0 \cdot \cosh(z/a_0)$, $|z| \leq h/2$. Since $h/2 = a_0 z_0$, the boundary of this catenoid, defined over $M_{h/2} = M_{a_0 z_0}$, is the only conjugate boundary of the exhaustion $\{M_t\}$. This proves that the bifurcation occurs on the conjugate boundary.

Now if we stretch two catenoids on the circles of radius ρ, and h is the distance between these circles, then the line $y = \rho$ intersects the graph $y = a\cosh(h/2a)$ in two points (a_1, ρ) and (a_2, ρ), $a_1 < a_2$. The equations of the corresponding catenoids have the form $r = a_1 \cosh(z/a_1)$ and $r = a_2 \cosh(z/a_2)$.

Since $h/2 = a_0 z_0$, each of these catenoids is defined over the domain $M_{h/2} = M_{a_0 z_0}$. For the first catenoid the critical domain of the exhaustion, that is, the only domain of the exhaustion $\{M_t\}$ having a conjugate boundary, is $M_{a_1 z_0}$, and for the second it is $M_{a_2 z_0}$. Since $a_1 z_0 < a_0 z_0 < a_2 z_0$, we have $M_{a_1 z_0} \subset M_{a_0 z_0}$ for the first catenoid and $M_{a_0 z_0} \subset M_{a_2 z_0}$ for the second. Therefore the first catenoid is unstable, while the second is stable.

APPENDIX

Steiner's Problem for Convex Boundaries

In Chapter 1 we spoke about Steiner's problem, which, we recall, consists in constructing a network of minimal length joining n given points in a plane. In this Appendix we present recent results of A. O. Ivanov and A. A. Tuzhilin [41], in which they obtained a classification of nondegenerate minimal networks without cycles and with convex boundary up to planar equivalence (Theorems 1 and 2). They discovered that all such minimal networks can be described as dual graphs to "planar tree tilings"; for the exact definition of these see below. They also produced an algorithm, realized on a computer, for calculating all networks of the given type for any fixed number of boundary points. Also, the authors found several infinite series of minimal networks with a regular convex boundary and individual interesting examples of such networks not contained in them. In the latter case it is important that the set of boundary points of the network is fixed, which considerably complicates the investigation.

1. General statement of the problem. We shall understand the minimality of a network in the following sense: any small fragment of the network has the shortest length. We recall that soap films have a similar property. When it is a question of shortest length, we need to distinguish the class of admissible variations of the network. First of all, by analogy with the way we defined minimal surfaces by means of the variational principle, we restrict the admissible variations just to those that leave fixed the initial points spanned by the network (henceforth we shall call these points the *fixed points* of the network). Two possibilities arise.

The first possibility: under a deformation of the network its vertices do not split, that is, variations of the type shown in Figure 48a are forbidden. In this case it can be shown (see [33], for example) that the network is minimal if and only if for each moving point the sum of the unit vectors having the directions of the segments going out from it is equal to zero. In this sense the network given in Figure 48a is minimal. In Figure 48b we give an example of a network that is not minimal in this sense.

The second possibility: The vertices of the network are allowed to split. In

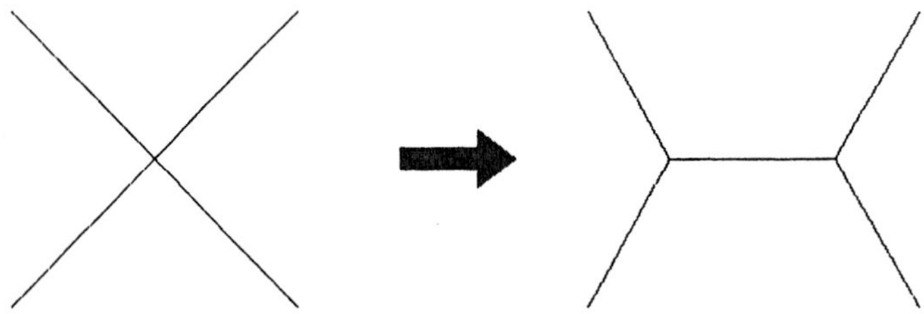

FIGURE 48A

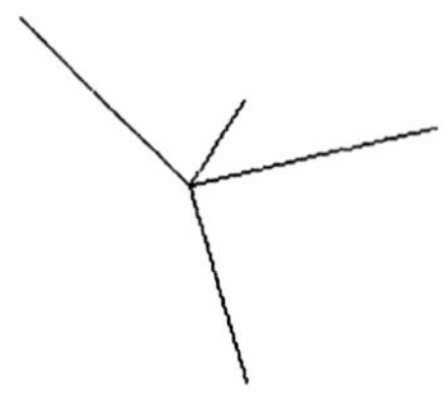

FIGURE 48B

this case under a deformation of the network in the direction of decreasing length its vertices split into points of degree at most three, and if three segments now meet at a vertex, then the angles between them are equal to 120°; if two segments meet at a vertex, then the angle between them is at least 120° and this vertex is fixed; also, each vertex from which exactly one segment goes out is fixed.

All these effects can be observed in the following simple experiment. We take a flat sheet of plexiglass and drill n small holes in it (Figure 49a). These holes will correspond to the fixed points of the network. From a piece of string we cut a set of $n-1$ segments. On one end of $n-2$ of the segments we make a small loose loop. We take the segment without a loop and pass it through an arbitrary number of loops. We can again pass the ends of the resulting configuration through a certain number of loops and so on, continuing this

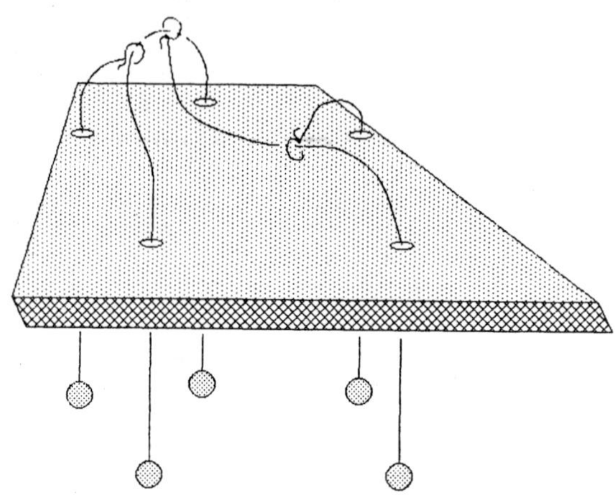

FIGURE 49A

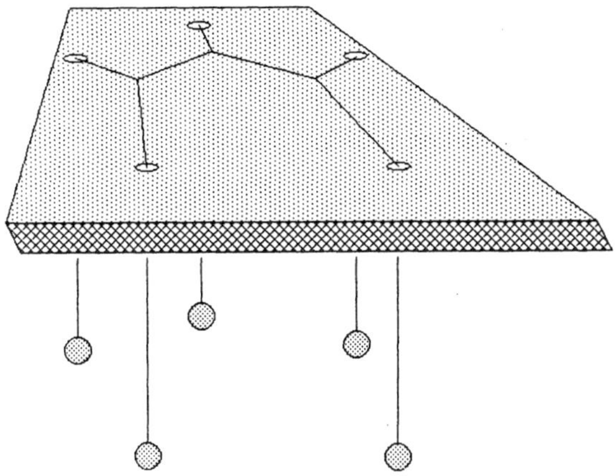

FIGURE 49B

process until all the segments have been used. We note that the number of ends of the resulting configuration is equal to the number of holes.

We place our sheet horizontally and pass all the ends of the resulting configuration upwards through the holes so that through each hole strictly one end passes. To each end we fasten a load, the loads being equal in mass.

After the system arrives at an equilibrium position, the network of string takes the form of a minimal network in one of the senses described above (Figure 49b). More concretely, if all the loops are separated, without interfering with one another, we obtain a minimal network in the second sense. But if at least one pair of loops is coupled, then there is a disallowed vertex in the resulting network, and the network is minimal only in the first sense.

Henceforth we shall study networks that are minimal in the second sense. In order to state our problem more precisely, we give the following definitions.

DEFINITION 1. A *topological Steiner network* is defined as a connected graph for which the degree of the vertices is at most three.

A realization of a topological network in a plane is called a *planar network.* To give a stricter definition we recall that a planar graph is a collection of curves in the plane that intersect only at their ends.

DEFINITION 2. A *planar Steiner network* is a planar graph for which there is a one-to-one correspondence with a topological Steiner network under which vertices correspond to vertices, curves correspond to edges, and the incidence relation is preserved.

A *set of fixed points* of a planar Steiner network is defined as an arbitrary subset of the set of vertices of the corresponding planar graph in which there occur all vertices of degree one or two (which agrees with the description of possible types of vertices of a network that is minimal in the second sense). Clearly, a set of fixed points of a network is not uniquely defined.

DEFINITION 3. A planar Steiner network is said to be *minimal* if it is minimal for some set of fixed points of it.

THE GENERAL STEINER PROBLEM. Describe the class of Steiner networks that can be realized as minimal networks. More precisely, suppose we are given a class $\{M\}$ of finite sets M of points of the plane. It is required to describe all Steiner networks that can be realized as minimal networks with a set of fixed points lying in this class.

REMARK. In Chapter 1 we talked about closed minimal networks on a sphere (we needed to study them to prove Plateau's principles). The problem of describing closed minimal networks on a closed two-dimensional orientable surface of genus g is an interesting generalization of Steiner's problem. A description of special classes of such networks was recently obtained by Shklyanko [34].

To start with, we consider as the class $\{M\}$ all possible subsets of points of the plane.

PROBLEM 1. Describe all Steiner networks that can be realized as minimal networks.

The class $\{M\}$ is the widest of all possible classes. However, it is not possible to realize all Steiner networks even on this. Figure 50a shows an example of a topological Steiner network that cannot be realized as a planar network, and hence as a minimal network. Figure 50b shows a planar Steiner network that cannot be realized as a minimal network. We note that in both cases all the trouble arises because of the presence of cycles. It turns out that this is the only obstacle to the realization of a Steiner network as a minimal network.

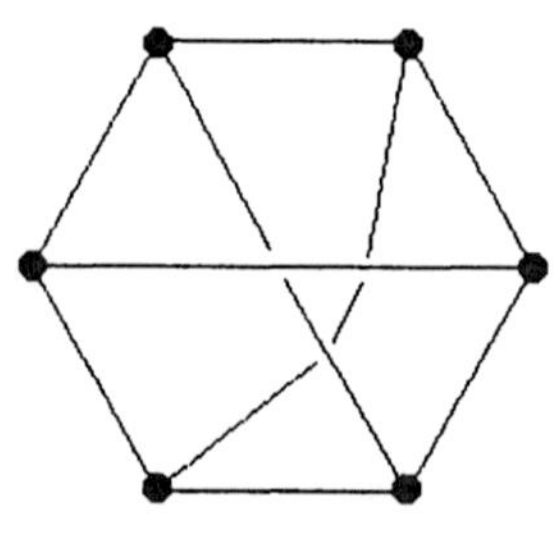

FIGURE 50A

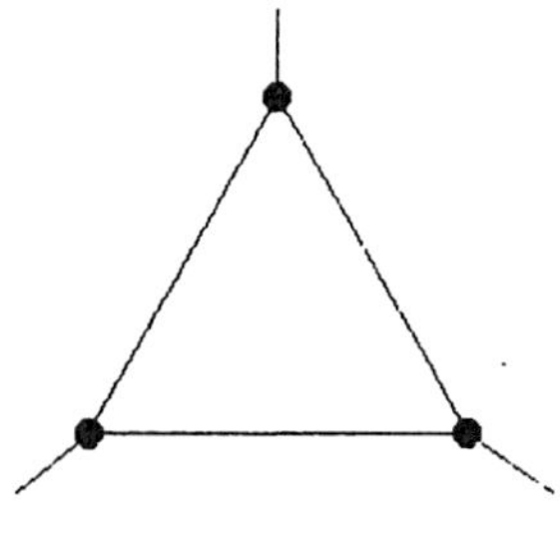

FIGURE 50B

PROPOSITION 1. *Any acyclic topological Steiner network (Steiner tree) can be realized as a planar network. Moreover, any planar Steiner tree can be realized as a minimal network for some set M of fixed points.*

We say that a Steiner network is *degenerate* if it has at least one vertex of degree two. We note that nondegenerate acyclic Steiner networks, by definition, are 2-trees. These networks have vertices of degree one, which we call *boundary vertices*, and vertices of degree three, which we call branch *points*. For such networks it is natural to take the boundary points as fixed, and we shall do this from now on. Starting from here we shall study acyclic nondegenerate minimal Steiner networks, that is, *minimal 2-trees*.

On the set of all planar graphs we can introduce a natural equivalence relation. We say that two planar graphs are *equivalent* if there is a homeomorphism of the plane onto itself (preserving the orientation) that takes one planar graph into the other. It is easy to see that there are only finitely many equivalence classes of 2-trees with a fixed number of boundary points.

The following natural question arises: how many such equivalence classes are there for a given number n of boundary points? From Proposition 1 it follows that there are exactly as many of them as there are equivalence classes of 2-trees. The number of the latter was calculated in 1964 by Brown [35] in implicit form. The numerical results for $n \leq 23$ can be found in [36].

We note that to solve this problem, instead of planar trees we can consider the dual objects, namely triangulations by diagonals of convex n-gons. Let us describe in more detail the correspondence between planar 2-trees and triangulations.

Suppose we are given a 2-tree with n boundary points. Consider a convex n-gon. We number the boundary points of the 2-tree in succession, going around anticlockwise, for example. Similarly we number the sides of the polygon. These numberings generate a natural one-to-one correspondence between the vertices of the 2-tree and the sides of the n-gon.

Obviously, the boundary points incident with the same branch point have consecutive numbers. For each pair of such points we consider the corresponding pair of sides of the n-gon and construct a triangle on these sides, drawing a diagonal of the polygon (this can always be done, since these sides are adjacent).

We cut out all the triangles obtained in this way and simultaneously discard from the 2-tree all the edges going out from the boundary points. Obviously, we again obtain a convex polygon and a 2-tree, and the number of vertices and the number of boundary points, respectively, are equal.

Between the boundary points and sides of the resulting objects there is a natural one-to-one correspondence, which is obtained directly from the correspondence established at the previous stage.

We repeat the procedure just described until the 2-tree is exhausted. (The last stage is a little more delicate, but it does not present any essential difficulty, so we leave the details to the reader.)

As a result we obtain a partition of the convex n-gon into triangles, which is called a *triangulation by diagonals* corresponding to a planar 2-tree.

Conversely, if we are given a triangulation of a convex n-gon by diagonals, it is easy to construct the corresponding 2-tree. As boundary points of such a tree we can take the midpoints of the sides of the n-gon, as branch points the centers of the triangles of the triangulation, and as edges the segments joining the centers of adjacent triangles and also the segments joining the midpoints of the sides of the n-gon to the centers of the triangles constructed on these sides.

Two triangulations are said to be *equivalent* if they are equivalent as planar graphs. In each equivalence class it is convenient to choose as a representative the corresponding triangulation of a regular polygon inscribed in the unit circle. Two triangulations of such regular polygons are equivalent if they are obtained from each other by a motion of the plane. It is easy to see that equivalent planar 2-trees correspond to equivalent triangulations of convex n-gons by diagonals and conversely.

Thus, the following proposition is true.

PROPOSITION 2. *The equivalence classes of planar 2-trees with n boundary points are in one-to-one correspondence with the equivalence classes of triangulations of convex n-gons by diagonals.*

Figure 51 shows all possible triangulations in the cases when $n = 3, 4, 5, 6$. We note that for $n < 6$ the triangulation by diagonals is unique (up to equivalence), but for $n = 6$ there are three different triangulations.

Another natural class $\{M\}$ of boundary points of networks is the class of extremal sets. We recall that a set is called *extremal* if it lies on the boundary of some convex set. If the set of boundary points of a network is extremal, such a network is called a *network with convex boundary*.

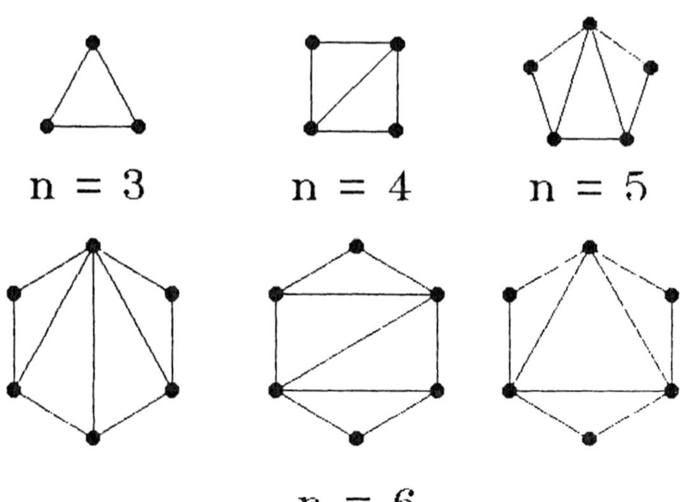

FIGURE 51

PROBLEM 2. Describe all minimal Steiner networks with convex boundary.

REMARK. This problem can be generalized. For this we need the following definition.

DEFINITION 4. Suppose we are given an arbitrary finite set M of points of the plane. We split it into classes, which we call *levels of convexity*.

In the first level of convexity we put all points lying on the boundary of the convex hull of M. Consider the set M' obtained from M by discarding all points of the first level.

The second level of convexity contains the points of the first level of convexity for the set M' (if M' is not empty).

Continuing this operation until all the original set M is exhausted, we obtain the necessary partition.

We observe that extremal sets, and only they, have exactly one level of convexity.

PROBLEM $2'$. Describe all minimal Steiner networks for which the number of levels of convexity of sets of boundary points does not exceed some fixed number.

Below we give a complete solution of Problem 2 for 2-trees.

One more important variant of the general Steiner problem is obtained if for the class $\{M\}$ of sets of boundary points of networks we consider the class consisting of exactly one set.

PROBLEM 3 (the classical Steiner problem). Describe all minimal Steiner networks whose set of boundary points is fixed.

An interesting variant of this problem is the following.

PROBLEM $3'$. Describe all minimal Steiner networks whose set of boundary points consists of the vertices of a regular polygon.

Below we give some results of A. O. Ivanov and A. A. Tuzhilin, devoted to investigations of Problem $3'$ again for 2-trees.

In connection with the statement of the general Steiner problem, the following interesting question arises: is there a set M consisting of n points on which all equivalence classes of planar 2-trees with n boundary points can be realized as minimal networks? For $n = 3, 4, 5$ we can take as such a set the vertices of the corresponding regular n-gon. For $n > 5$ this is not so. From Proposition 3 (see below) it follows that, generally speaking, such a set M must have quite a complicated structure (for example, it cannot be extremal).

2. Classification of minimal 2-trees with convex boundary. An important role in the classification of minimal 2-trees with a convex set of boundary points is played by the so-called twisting number; we begin this section with a definition of it.

For each branch point of a planar 2-tree there is a circular neighborhood whose intersection with the tree consists of three smooth nonclosed curves going from its center, not having any other points of intersection, and going out to the boundary of this neighborhood. Clearly, the intersection of the boundary of the neighborhood (the circle) with the tree consists of three points. We choose one of these curves and call the edge, of which it is a part, *incoming*. We call the remaining two edges *outgoing*. We call the partition of the edges incident with a point into incoming and outgoing an *orientation* of the neighborhood of the branch point.

Now suppose that the plane is oriented. Then we are given a positive direction for motion along each circle lying in this plane. This gives the possibility of naturally ordering a pair of outgoing edges in such a way that a motion along the circle from the first edge to the second along an arc that does not intersect an incoming edge takes place in the positive direction. We assign the number -1 to the first outgoing edge, and $+1$ to the second. An oriented neighborhood of a branch point together with these numbers is said to be *clothed*, and the numbers themselves are called *clothings* of the corresponding edges of the tree.

We recall that a path joining a pair of edges of a planar 2-tree is defined as a minimal connected subtree containing these edges. We now give a definition of the twisting number between a pair of edges of a planar 2-tree.

Let a and b be a pair of edges of a planar 2-tree. We choose a path γ joining a and b. We orient the path γ from a to b. Consider all branch points lying inside γ. The orientation of γ canonically specifies orientations of small neighborhoods of these branch points; for each point we shall take as incoming an edge for which the point is an end. We fix a certain orientation of the plane and clothe the neighborhoods of all the branch points under consideration. A path γ together with clothings of all its edges will be called *clothed*.

DEFINITION 5. The *twisting number* $\mathrm{tw}(a, b)$ of an ordered pair (a, b) of distinct edges of a 2-tree is defined as the sum of the clothings of all outgoing edges of the oriented path γ going from a to b. We take $\mathrm{tw}(a, a)$ to be zero.

For the edges a and b of the 2-tree shown in Figure 52a the twisting number $\mathrm{tw}(a, b)$ is five, while for the tree in Figure 52b it is zero.

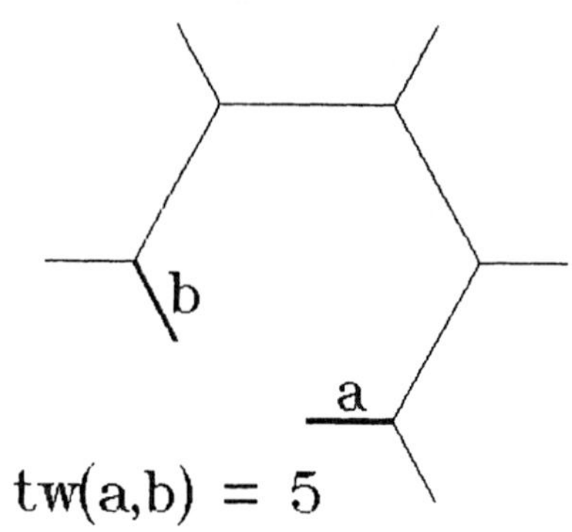

FIGURE 52A

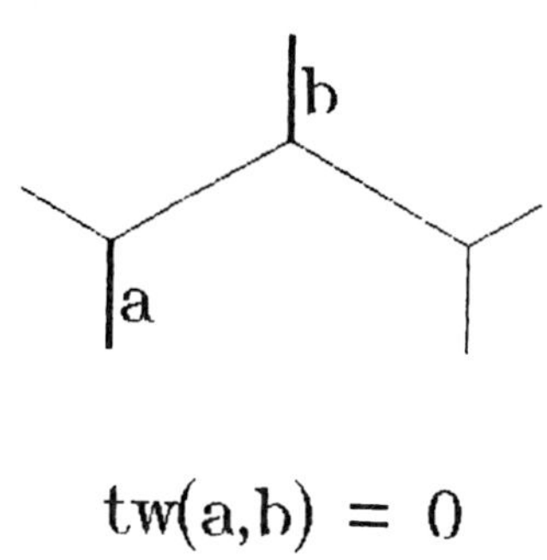

FIGURE 52B

Let us mention a property that the twisting number has (we leave the proof to the reader as a useful exercise):

SKEW-SYMMETRY: $\mathrm{tw}(a, b) = -\mathrm{tw}(b, a)$.

DEFINITION 6. The *twisting number* $\mathrm{tw}(D)$ of a planar 2-tree D is defined as the largest twisting number of all possible ordered pairs of edges of this tree:

$$\mathrm{tw}(D) = \max \mathrm{tw}(a, b).$$

The next proposition is a key result in obtaining a complete classification of minimal 2-trees with convex boundary.

PROPOSITION 3. *The twisting number of a minimal Steiner 2-tree with convex boundary is not greater than five.*

REMARK. From the classification theorems (see below) it follows that this bound is exact: any planar 2-tree with twisting number not greater than five can be realized as a minimal tree with an extremal set of boundary points.

It is convenient to state the classification theorem in the language of tilings. We define a (triangular) tiling of the plane as a canonical partition of the plane into regular congruent triangles, which we call the *cells* of the tiling. This partition can be obtained as follows.

Let A and B be families of equally spaced parallel lines, and suppose that the angle between the directions of the lines of A and B is 60°. Through the points of intersection of lines of A and B we can uniquely draw a third family C of parallel lines so that together the families A, B, and C give a partition of the plane into regular congruent triangles. We call these lines the *directrices* of the tiling of the plane, and the six possible directions of these lines the *directions* of the tiling.

DEFINITION 7. We define a *tiling* as an arbitrary collection of cells of a tiling of the plane.

In exactly the same way as from a triangulation of a convex polygon by diagonals, from the tiling we can construct a planar graph, which we call the *dual graph* of this tiling. We call a tiling *connected* if its dual graph is connected. The tilings corresponding to the connected components of the dual graph are called the *components* of the tiling. Henceforth we shall almost always be dealing with connected tilings, so we shall omit the word "connected" provided it does not lead to misunderstanding.

We note that the dual graph of an arbitrary (connected) tiling is actually a minimal Steiner network.

DEFINITION 8. A tiling whose dual graph is a 2-tree is called a *tree tiling*.

In fact, not every equivalence class of planar 2-trees has as its representative the dual graph of some tree tiling. Nevertheless, the following proposition is true.

PROPOSITION 4 (on a tiling realization). *Any planar 2-tree with twisting number no greater than five can be realized as the dual graph of some tree tiling.*

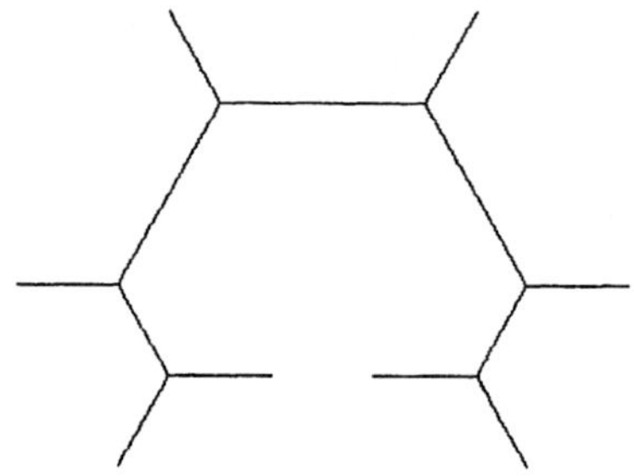

FIGURE 53

REMARK. Although some 2-trees with twisting number greater than five can be realized as the dual graph of a tree tiling (construct an example), the bound on the twisting number given in Proposition 4 is exact. Figure 53 shows an example of a planar 2-tree with twisting number equal to six that cannot be realized as the dual graph of a tree tiling.

Thus, it follows from Propositions 3 and 4 that for a classification of minimal 2-trees with convex boundary it is sufficient to describe all tree tilings whose dual graphs have twisting number not exceeding five (henceforth for brevity we shall call the twisting number of the dual graph of a tree tiling the twisting number of the tiling itself).

To obtain such a classification we must first of all choose the building blocks from which all possible tree tilings are formed. We choose three types of building blocks, which we shall call *linear parts, branch points*, and *growths*. Roughly speaking, every tree tiling is a collection of linear parts joined to one another by means of branch points and equipped with growths. We now give more formal definitions.

DEFINITION 9. We define a *snake* as a tiling placed between two adjacent directrices of a tiling of the plane (Figure 54).

An *extreme cell* is defined as a cell of a tiling, two sides of which do not lie inside the tiling. An *interior cell* is defined as a cell, all of whose sides lie inside the tiling.

DEFINITION 10. We call an extreme cell of a tiling a *growth* if the only cell

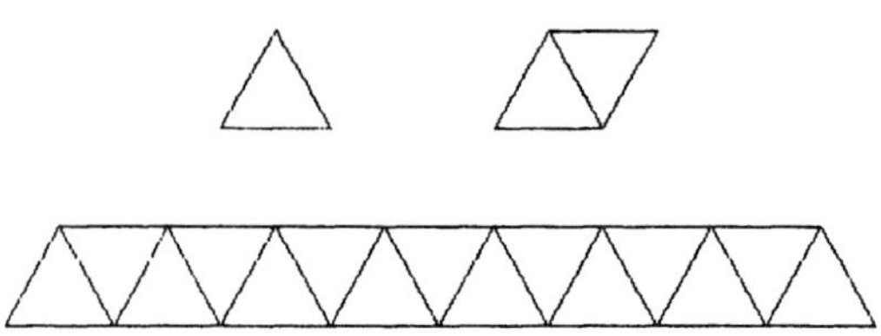

FIGURE 54

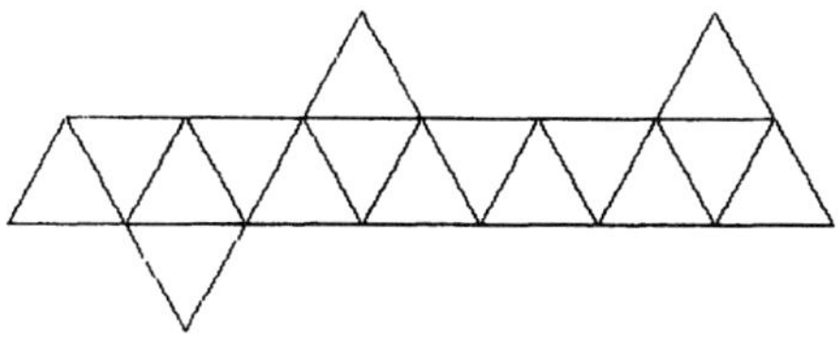

FIGURE 55A

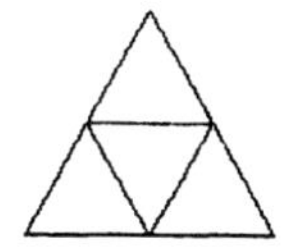

FIGURE 55B

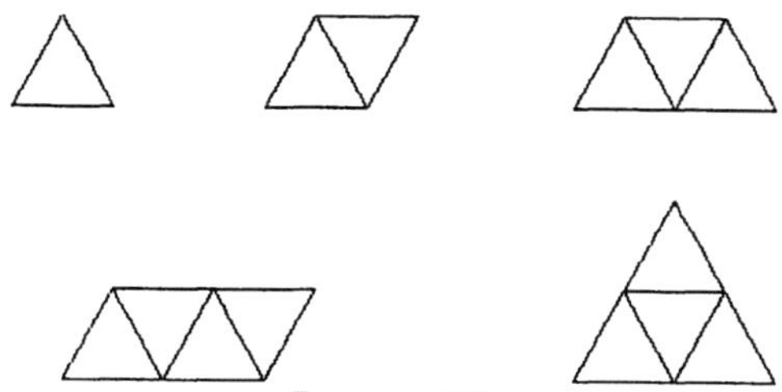

FIGURE 56

of the tiling adjacent to it is internal. We call the tilings a *skeleton* if it does not have growths.

Figure 55 shows a snake with growths.

In order to obtain a skeleton, for each interior cell of the tiling we discard one of the growths adjoining it (if there are any).

Next, we consider a tree skeleton and cut out from it all internal cells. If there is at least one internal cell, the skeleton splits into components.

DEFINITION 11. The components into which a tree skeleton splits after discarding internal cells are called the *linear parts* of the skeleton. The *branch points* are the components into which a tree skeleton splits after discarding the linear parts.

Let us give a complete list of possible branch points of tree skeletons.

PROPOSITION 5. *In tree skeletons there can occur exactly five types of branch points, shown in Figure* 56.

REMARK. We should mention that the linear parts can be fastened to each branch point in different ways. In all there are 18 ways of fastening (list them), which we shall call *forks*. Figure 57 shows the two most important types of forks, which we shall call *T-joints*.

We now describe the structure of the linear parts. For this we give the more general definitions of a linear 2-tree and a linear tiling.

DEFINITION 12. A planar 2-tree is called *linear* if the triangulation of a convex polygon corresponding to it has exactly two extreme triangles (an extreme triangle of a triangulation is a triangle of which two sides coincide with sides of the polygon).

We note that the triangulation corresponding to a linear 2-tree does not have internal triangles. Therefore, for such a triangulation there is a natural linear ordering of its triangles so that the extreme triangles are the first and last in this order.

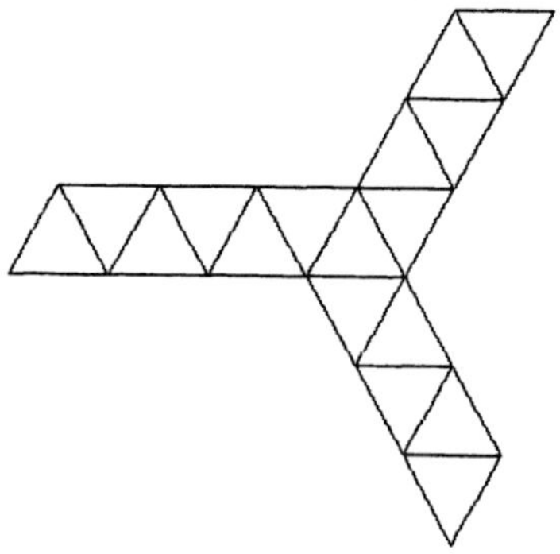

FIGURE 57A

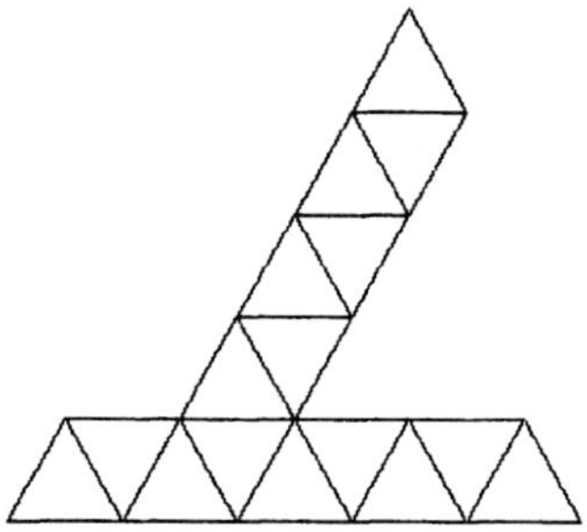

FIGURE 57B

Clearly, there are exactly two possibilities for ordering the triangles of a triangulation, depending on which of the two extreme triangles is taken as the first. The choice of one of these two orders is called an *orientation* of the linear 2-tree and the triangulation corresponding to it.

Now suppose that the dual tree of some tree tiling is linear. In this case the tiling is also said to be linear. We note that the linear parts are actually linear tilings.

We now consider an arbitrary linear tiling and orient it. We then show how we can split such a tiling into a number of snakes, which we call segments of the linear tiling.

Similarly we define all the later segments. Thus, we can state the following result.

PROPOSITION 6. *Every linear tiling, in particular, the linear part of an arbitrary tree tiling, is the union of a linearly ordered family of distinct disjoint snakes (segments of the linear tiling), and the initial cell of each subsequent snake is adjacent to the end cell of the previous one.*

In each nonextreme cell of a skeleton we join by segments the midpoints of its sides inside the skeleton. In each extreme cell we draw a midline parallel to the midline of the adjacent cell already constructed.

DEFINITION 13. The *spine* (*vertebra*) of a linear part (cell) is the part of the graph constructed above that is contained in this linear part (cell).

If we allow the twisting number of the skeleton to take only values not exceeding five, then there arise essential restrictions on the structure of the linear parts of such a skeleton. Namely, the following proposition is true.

PROPOSITION 7. *For each linear part of a tree skelelton with twisting number not exceeding five there is a directrix of the tiling of the plane on which the spine of this linear part projects one-to-one. Such a directrix is called a directrix of the linear part.*

REMARK. Generally speaking, a directrix of a linear part is not unique. For a snake, for example, there are three such directrices. If the twisting number of the linear part is greater than five, then such a part does not have directrices.

DEFINITION 14. A snake is defined as a linear part for which there are three directrices. A *stairs* is defined as a linear part for which there are exactly two directrices. A linear part that has exactly one directrix is called a *broken snake* (Figure 58a).

REMARK. A linear part that is a snake in the sense of Definition 9 may not be a snake in the sense of the last definition. Figure 58b shows an example of such a part. Of course, it all depends on how the given linear part is fastened to the branch points.

We have thus described all the building blocks from which all possible tree tilings are formed. Now, in order to state the classification theorems, it remains to define the operation of reduction for skeletons of tree tilings with twisting number not exceeding five. This operation consists in cutting out certain fragments of the tiling.

Firstly, we can cut from the skeleton any part of a linear part containing an extreme cell.

Secondly, inside the skeleton we can discard any snake Z consisting of an even number of cells and occurring in some linear part. We observe that the snake Z is a parallelogram. Let us consider the pair of sides of this parallelogram not parallel to the spine of this snake. We shall call these sides the *bounding edges* of the snake Z. It turns out that the following assertion is true.

ASSERTION. *Let Z be an arbitrary snake consisting of an even number of cells and occurring in some linear part of the skeleton D with twisting number*

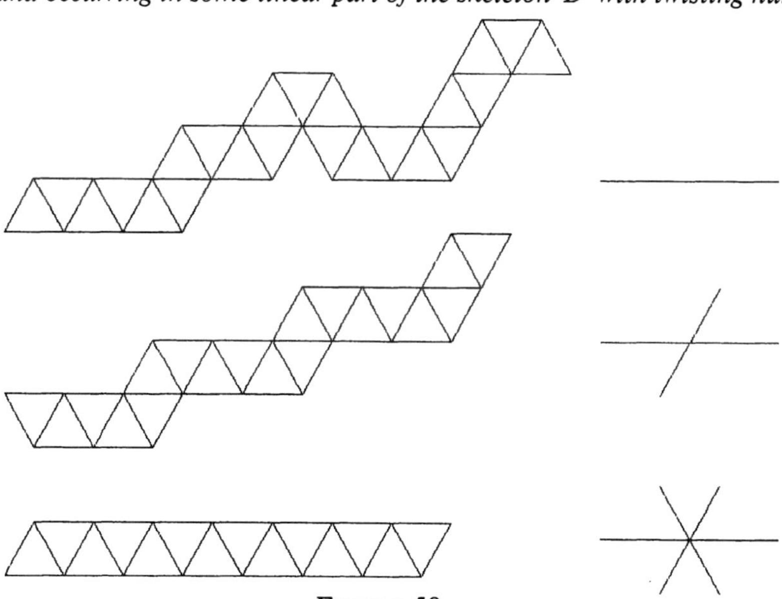

FIGURE 58A

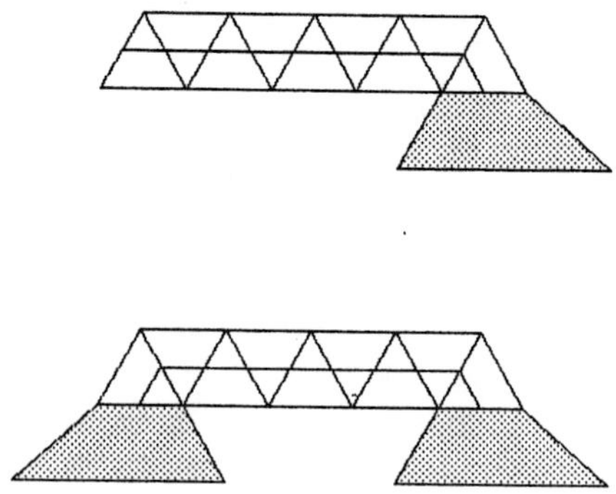

FIGURE 58B

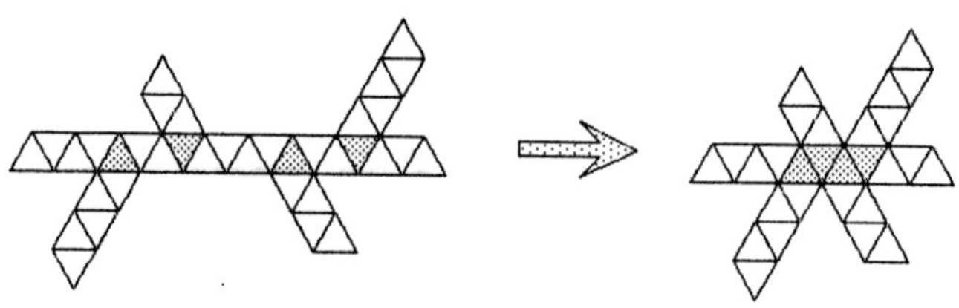

FIGURE 59

no greater than five. Let l_1 and l_2 be the bounding edges of Z. Let D_1 and D_2 be the connected components into which D splits after discarding Z from it. Then there is a translation τ such that the intersection of D_1 and $\tau(D_2)$ is $l_1 = \tau(l_2)$. Moreover, $D_1 \cup \tau(D_2)$ is a tree skeleton, whose twisting number is not greater than the twisting number of D.

We say that the skeleton $D_1 \cup \tau(D_2)$ is *reduced* from the skeleton D by cutting out the snake Z.

REMARK. By means of the reduction operation we can obtain a stairs from a broken snake and a snake from a stairs. By reduction we can turn several branch points into one point (of a different type). This also happens with forks. Figure 59 shows the reduction of several forks of T-joint type to a fork of a more complicated form. The reduction operation also enables us to discard forks (we need to apply reduction several times).

We are now in a position to state a theorem that classifies skeletons of tree tilings with twisting number not exceeding five.

THEOREM 1 (classification of skeletons) (Ivanov, Tuzhilin). *All skeletons with twisting number not exceeding five are obtained by reduction from the three canonical types of skeletons given in Figure 60.*

A broken snake is represented by three dashes, and the dashes are parallel to its directrix.

A stairs is represented by two intersecting dashes, and the dashes are parallel to its two directrices.

A snake is represented by one dash, and the dash is parallel to the spine of the snake.

Forks of T-joint type correspond to points (see Figure 57).

REMARK. If we consider the diagrams of the canonical types as planar 2-trees, it is easy to observe that they represent all possible planar 2-trees with six endpoints.

We now describe the possible positions of growths on a skeleton. For this we need the concept of a profile of a skeleton.

The *contour* of a tiling is defined as the boundary of the tiling regarded as a closed subdomain of the plane. Consider an extreme cell of the skeleton and discard from the contour of the skeleton that edge of it that intersects the vertebra of this cell. We go through all the extreme cells of the skeleton,

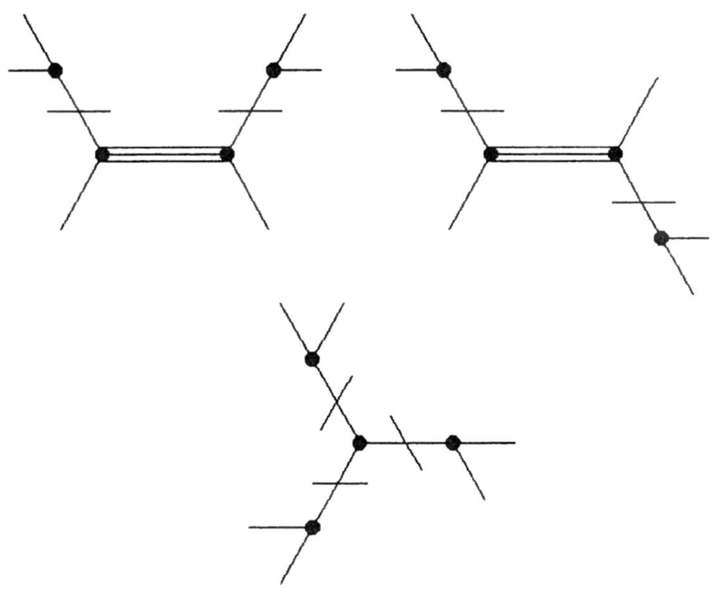

FIGURE 60

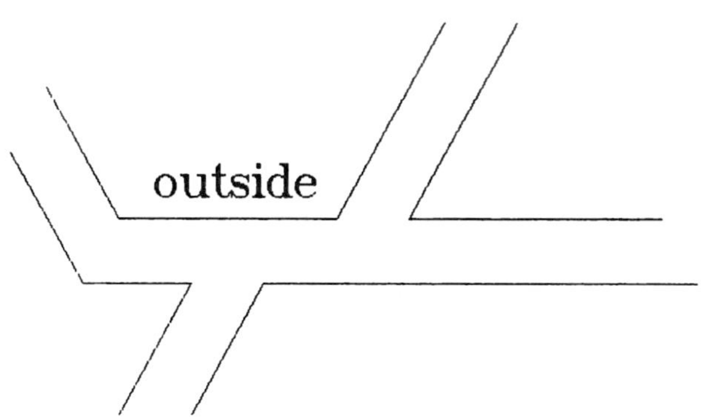

FIGURE 61

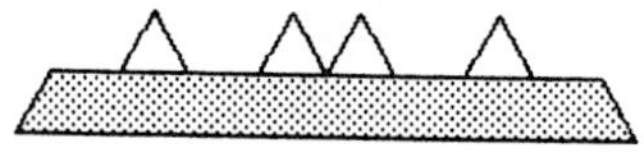

FIGURE 62A

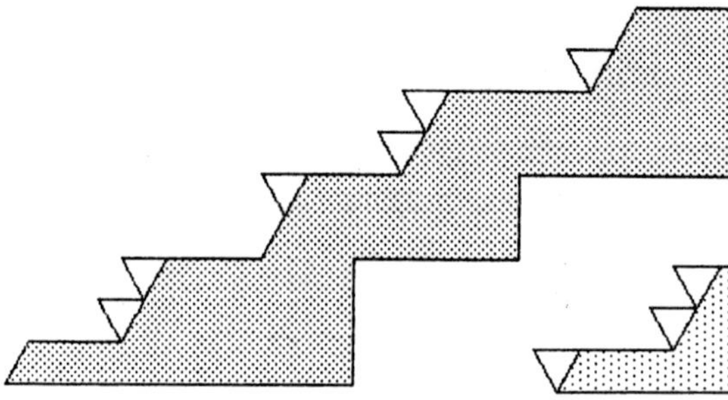

FIGURE 62B

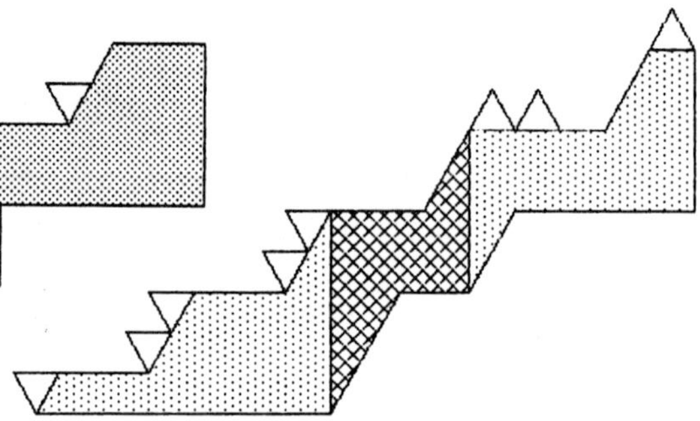

FIGURE 62C

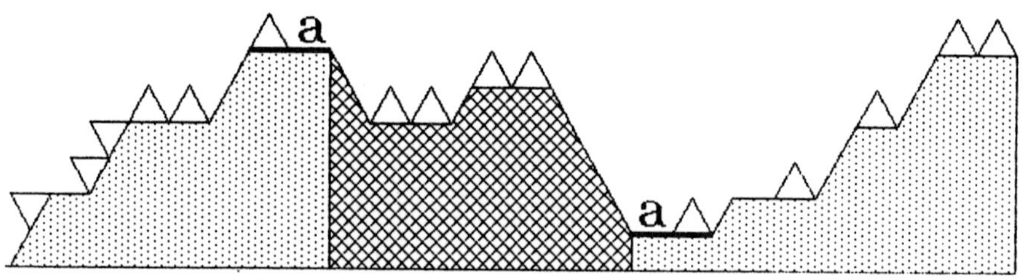

FIGURE 62D FIGURE 62E

performing the same operation. The contour of the skeleton splits into broken lines, which we call the *profiles* of the skeleton. An outer side of a profile is called an outer side of it with respect to the skeleton (Figure 61).

We note that the profiles of the skeleton of a tiling with twisting number not exceeding five have the same properties as the spines of the linear parts of such skeletons. Therefore for the corresponding profiles we keep the names snake, stairs, and broken snake.

THEOREM 2 (on the position of growths). (Ivanov, Tuzhilin) 1. *On a profile that is a snake we can plant any number of growths* (*Figure* 62a).

2. *For a stairs-profile there are two possibilities.*

a) *The growths are placed arbitrarily only on segments in one direction* (*Figure* 62b).

b) *We are given a partition of the stairs into three successive broken lines, the middle one of which may be empty. The middle broken line consists of an even number of links and the angle between the first pair of links, measured from the outer side, is equal to* 120°.

There are no growths on the middle broken line. On the first broken line the growths can be situated arbitrarily on the segments that have the direction of its last link, and on the last broken line they can be situated on segments having the direction of its first link (*Figure* 62c).

3. *We present a profile that is a broken snake as the union of three parts, the outer ones of which are maximal possible stairs, and the inner one, which may be empty, is all the rest. On the middle part we can plant arbitrarily many growths only on segments parallel to the directrix of the profile. On the outer stairss we can plant growths as follows.*

Consider a segment of the profile adjacent to an outer stairs. If the angle between it and the neighboring segment a *of the stairs, measured from the outer side of the profile, is equal to* $120°$, *then growths may be fastened only on segments of the stairs parallel to the segment* a. *If this angle is equal to* $240°$, *then we can plant growths on the stairs according to rule* 2 (*Figure* 62d, e).

Now the classification of possible minimal 2-trees with a convex set of boundary points is obtained from Propositions 3 and 4 and Theorems 1 and 2.

We can show that any planar 2-tree that is the dual graph to the tilings described in Theorems 1 and 2 can be realized as a minimal tree with a convex set of boundary points. Thus, the resulting classification is complete.

3. Some results from the investigation of minimal networks that span the vertices of regular polygons. We begin this section with a description of a simple algorithm: for a given finite set M of points of the plane this algorithm enables us to construct a minimal network spanning it by means of compasses and a straight edge (the idea of this algorithm is due to Melzak [42]). For this it is sufficient to know the structure of this minimal network as a planar 2-tree and the correspondence between the endpoints of this 2-tree and points of the set M. Recall, that the vertices of a minimal network that do not belong to M are called *Steiner points*. Let us illustrate the idea behind this algorithm by an example of constructing the minimal network for the set M of vertices of a triangle ABC, none of whose angles exceeds $120°$.

We choose any pair of vertices of the triangle, say A and B, and construct an equilateral triangle ABD on the side AB so that C and D lie on opposite sides of AB. We then describe the circle ABD.

Clearly, the only Steiner point V of the minimal network lies on the minor arc d of this circle joining A and B. Moreover, V lies on the ray DC (prove this). Joining V to the vertices of the triangle ABC, we obtain our minimal network (Figure 63a).

If the triangle ABC has an angle greater than or equal to $120°$, then the corresponding minimal network is not a 2-tree. In this case we can carry out the same construction, but the angles between the segments joining the point V of intersection of the ray DC and the circle to the vertices of the triangle will not be equal.

For a quadrilateral $ABCD$ the construction consists of two similar steps. Figure 63b shows a minimal network spanning the vertices of a square. We split the vertices of the square into pairs consisting of bounding vertices of

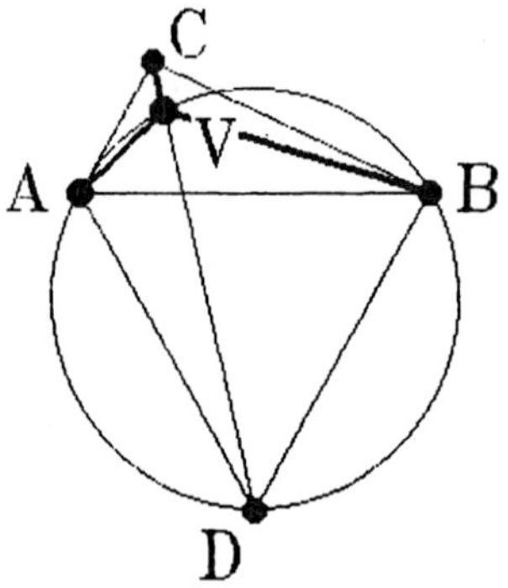

FIGURE 63A

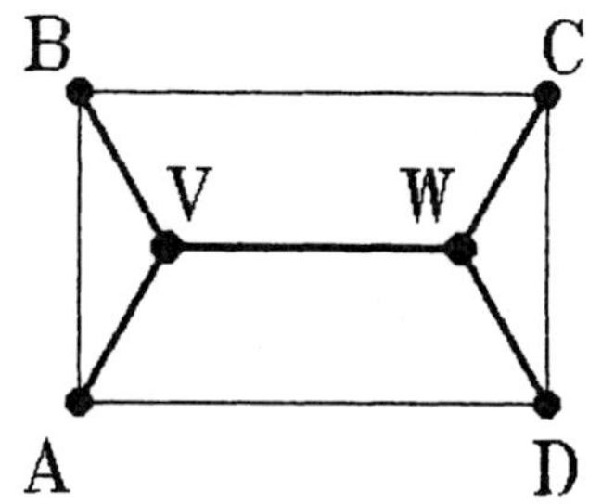

FIGURE 63B

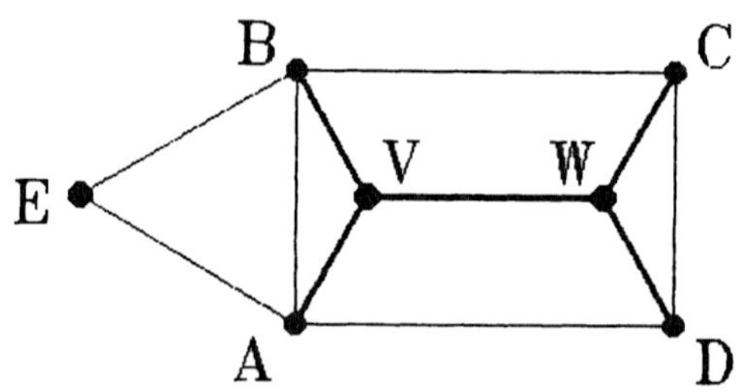

FIGURE 63C

the network for which the edges of the network going out from them meet at one Steiner point, and choose one of these pairs, say A and B. We denote by V the Steiner point at which the edges of the minimal network going out from A and B meet, and the other Steiner point by W.

On AB we construct an equilateral triangle ABE. We place the vertex E of this triangle in such a way that E and V lie on opposite sides of AB (Figure 63c).

We now consider the triangle CDE and describe the minimal network for it in the way described above. The Steiner point of this network coincides with W.

A minimal network for the vertices of the square $ABCD$ is obtained as follows. We describe the circle ABE. The point of intersection of this circle with the minimal network we have constructed is a Steiner point V of the required network (prove this). It remains to join V to A and B.

These ideas are the basis of the algorithm for constructing minimal networks with a given set of boundary points. This algorithm has been realized on a computer. For lack of space we do not give a detailed description of this algorithm here. Figure 64 gives minimal trees constructed by the computer.

A computer experiment has enabled us to formulate a number of conjectures about the structure of minimal 2-trees spanning the vertices of regular n-gons. Some of these conjectures have been proved. We give here a small part of the results we have obtained.

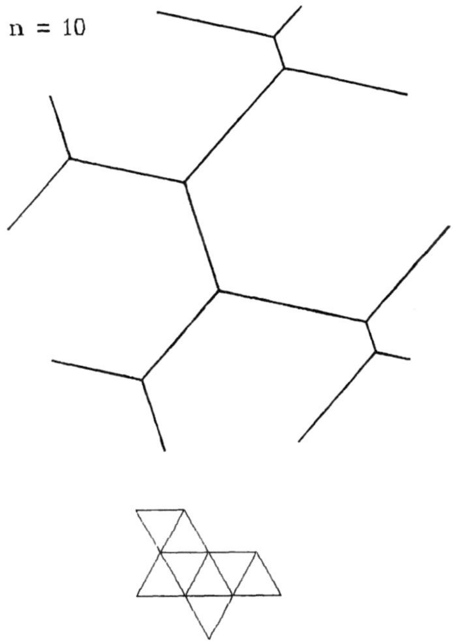

FIGURE 64A

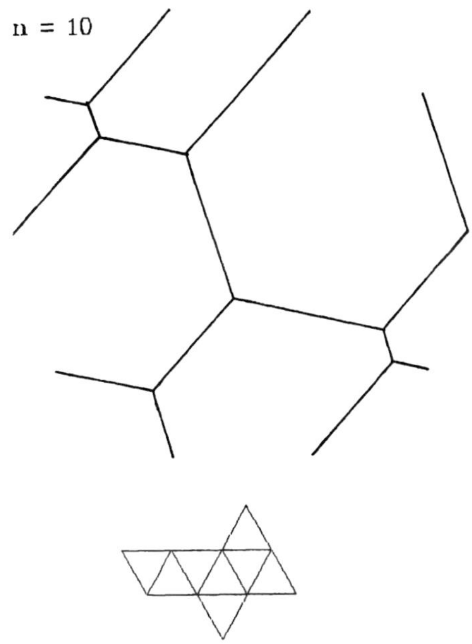

FIGURE 64B

PROPOSITION 8. *For any n, on the vertices of a regular n-gon we can stretch a minimal 2-tree of snake type uniquely up to a motion (Figure 65).*

PROPOSITION 9. *For any $n = 6k + 3$, where $k > 0$, on the vertices of a regular n-gon we can stretch a minimal 2-tree of T-joint type (from Figure 57a) uniquely up to a motion (Figure 66). This network is invariant under rotation about the center of the n-gon through 120°.*

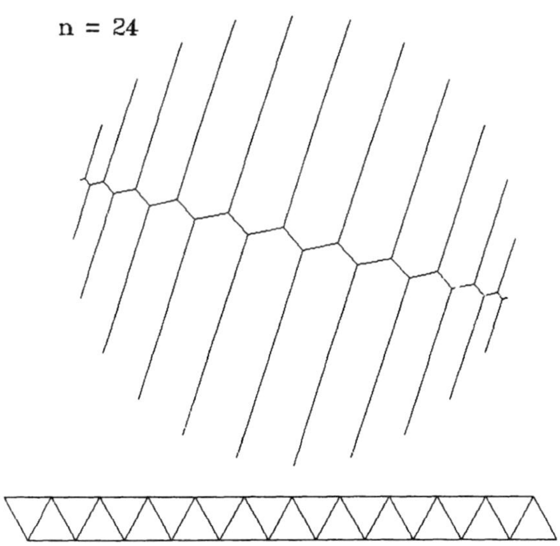

FIGURE 65

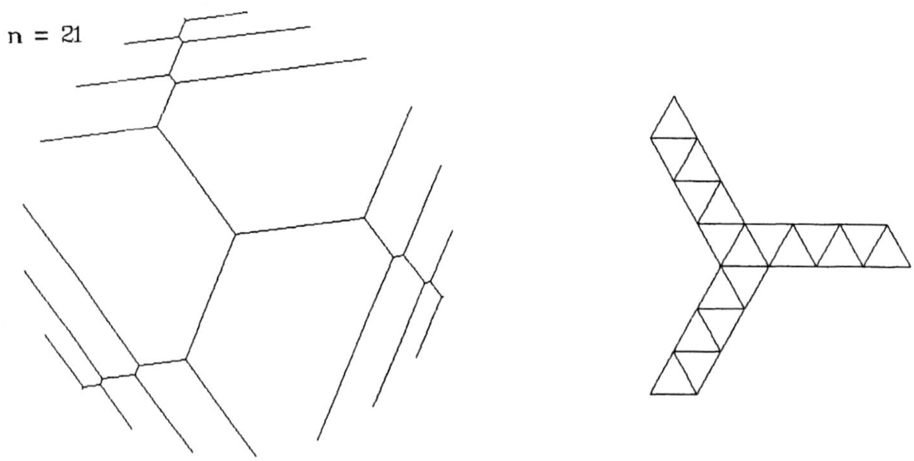

FIGURE 66

There is at least one infinite series of minimal trees—these are snakes with pairs of symmetrical growths situated close to the center of the snake (Figure 67). The authors have obtained estimates for the possible position of these growths, which we cannot give for lack of space.

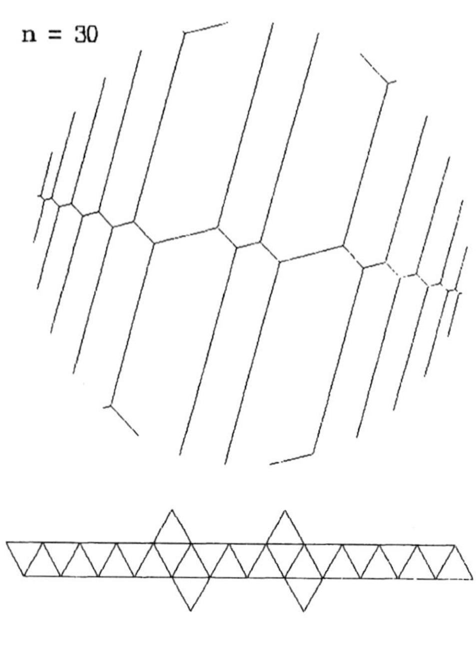

FIGURE 67

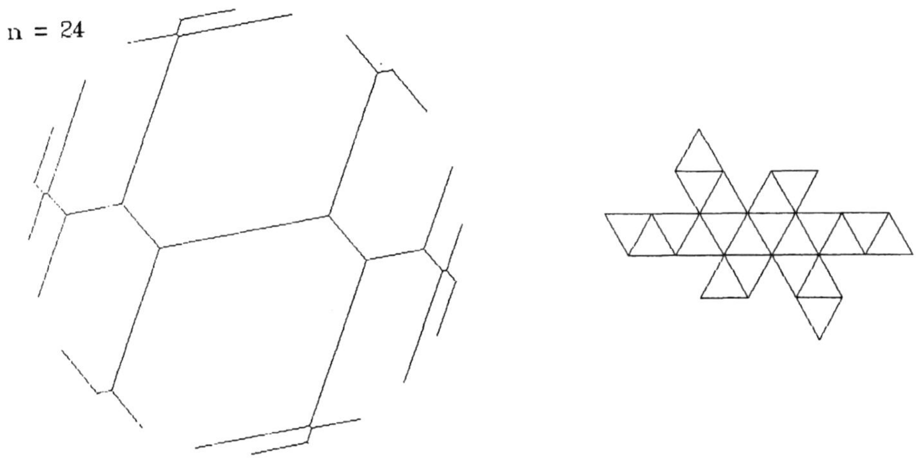

FIGURE 68A

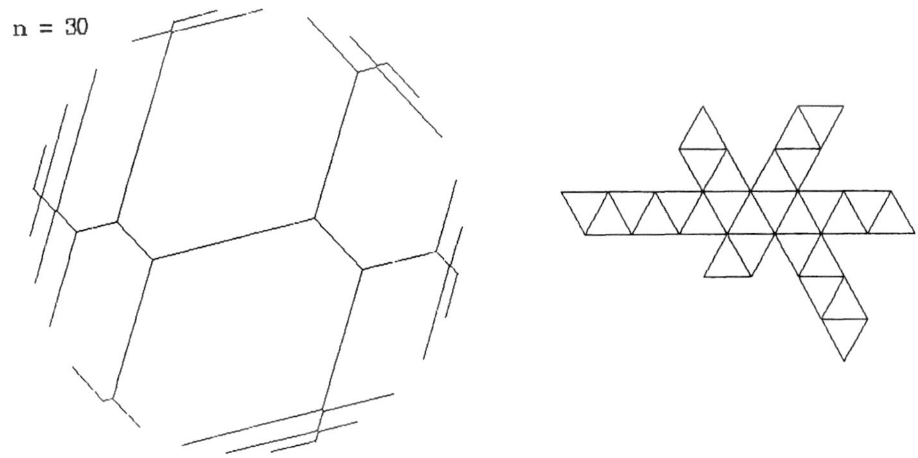

FIGURE 68B

Figure 68 gives representatives of an apparently finite series (as a computer experiment shows) of minimal trees realized on n-gons when $n = 24, 30, 36, 42$. We note that since the corresponding tilings have one branch point and six ends, there can be no growths on these networks.

Figure 69 gives an example of a network whose corresponding tiling has one branch point, four ends, and one growth. A computer experiment shows that there may exist an infinite series of such minimal trees.

These examples show that the problem of classifying minimal 2-trees whose sets of boundary points consist of the vertices of regular polygons is nontrivial.

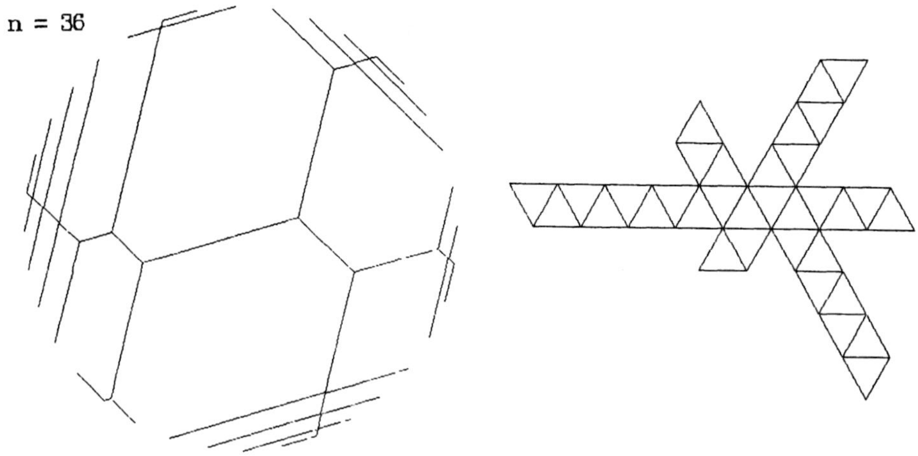

FIGURE 68C

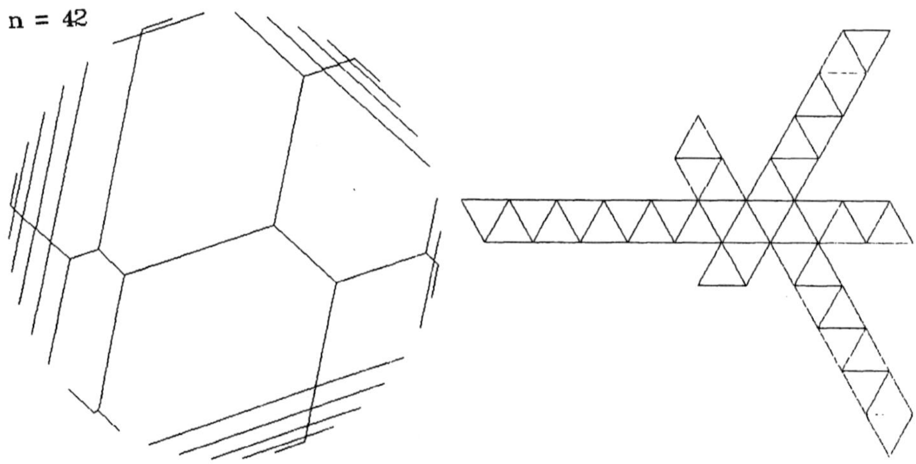

FIGURE 68D

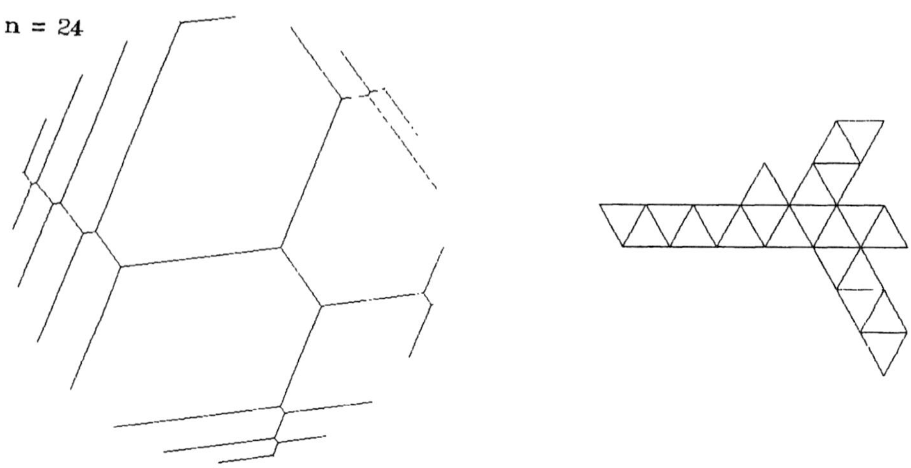

FIGURE 69

Bibliography

1. B. A. Dubrovin, S. P. Novikov, and A. T. Fomenko, *Modern geometry: methods and applications*, 2nd ed., "Nauka", Moscow, 1986. (Russian), English transl. Springer-Verlag, Part 1, GTM **93**, 1984; Part 2, GTM **104**, 1985; Part 3, GTM **104**, 1990.

2. Dào Trong Thi and A. T. Fomenko, *Minimal surfaces and Plateau's problem*, "Nauka", Moscow, 1987; English transl., Amer. Math. Soc., Providence, RI, 1991.

3. S. Hildebrandt and A. Tromba, *Mathematics and optimal form*, Scientific American Library, New York, 1985.

4. A. A. Tuzhilin and A. T. Fomenko, *Multivalued mappings, minimal surfaces and soap films*, Vestnik Moskov. Univ. Ser. Mat. **1986**, no. 3, 3–12; English transl. in Moscow Univ. Math. Bull. **41** (1986), no. 3.

5. R. Schoen, *Uniqueness, symmetry and embeddedness of minimal surfaces*, J. Differential Geom. **18** (1983), 791–809.

6. F. Morgan, *A smooth curve in* $\mathbb{R}^3$ *bounding a continuum of minimal manifolds*, Arch. Rational Mech. Anal. **75** (1981), 193–197.

7. S. Kobayashi and K. Nomizu, *Foundations of differential geometry*, Vol. 1, Interscience, New York-London, 1963.

8. I. N. Vekua, *Generalized analytic functions*, 2nd ed., "Nauka", Moscow, 1988. (Russian)

9. R. Osserman, *Minimal surfaces*, Uspekhi Mat. Nauk **22** (1967), no. 4, 55–136.

10. W. H. Meeks, III, *The classification of complete minimal surfaces in* $\mathbb{R}^3$ *with total curvature greater than* -8π, Duke Math. J. **48** (1981), 523–535.

11. M. Elisa and G. G. d'Oliveira, *Some new examples of nonoriented minimal surfaces*, Proc. Amer. Math. Soc. **98** (1986), 629–636.

12. R. Kusner, *Conformal geometry and complete minimal surfaces*, Bull. Amer. Math. Soc. (N.S.) **17** (1987), 291–295.

13. T. Klotz and L. Sario, *Existence of complete minimal surfaces of arbitrary connectivity and genus*, Proc. Nat. Acad. Sci. USA **54** (1965), 42–44.

14. L. P. Jorge and W. H. Meeks, III, *The topology of complete minimal surfaces of finite total Gaussian curvature*, Topology **22** (1983), 203–221.

15. J. A. Thorpe, *Elementary topics in differential geometry*, Springer, New York, 1979.

16. D. Hoffman, *The computer-aided discovery of new embedded minimal surfaces*, Math. Intelligencer **9** (1987), no. 3, 8–21.

17. A. I. Pluzhnikov, *Some properties of harmonic mappings in the case of spheres and Lie groups*, Dokl. Akad. Nauk SSSR **268** (1983), 1300–1302; English transl. in Soviet Math. Dokl. **27** (1983).

18. A. V. Tyrin, *On the absence of local minima in the multi-dimensional Dirichlet functional*, Uspekhi Mat. Nauk **39** (1984), no. 2, 193–194; English transl. in Russian Math. Surveys **39** (1984), no. 2.

19. M. Beeson and A. Tromba, *The cusp catastrophe of Thom in bifurcation of minimal surfaces*, Manuscripta Math. **46** (1984), 273–307.

20. J. L. Barbosa and M. Do Carmo, *On the size of a stable minimal surface in* $\mathbb{R}^3$, Amer. J. Math. **98** (1976), 515–528.

21. M. Do Carmo and C. K. Peng, *The stable minimal surfaces in* $\mathbb{R}^3$ *are planes*, Bull. Amer. Math. Soc. (N.S.) **1** (1979), 903–906.

22. D. Fischer-Colbrie, *On complete minimal surfaces with finite Morse index in three-manifolds*, Invent. Math. **82** (1985), 121–132.

23. J. Tysk, *Eigenvalue estimates with applications to minimal surfaces*, Pacific J. Math. **128** (1987), 361–366.

24. F. Lopez and A. Ros, *Complete minimal surfaces with index one and stable constant mean curvature surfaces*, Preprint, Granada, 1987.

25. T. M. Rassias, *Foundations of global nonlinear analysis*, Teubner, Leipzig, 1986.

26. J. Simons, *Minimal varieties in Riemannian manifolds*, Ann. of Math. (2) **88** (1968), 62–105.

27. S. Smale, *On the Morse index theorem*, J. Math. Mech. **14** (1965), 1049–1055.

28. A. A. Tuzhilin, *Indices of two-dimensional minimal surfaces*, Geometry and the Theory of Singularities in Nonlinear Equations, Voronezh Univ. Press, Voronezh, 1987, pp. 70–76. (Russian)

29. ——, *Indices of two-dimensional minimal surfaces in* $\mathbb{R}^3$, Selected Questions of Algebra, Geometry and Discrete Mathematics, Moscow Univ. Press, Moscow, 1988, pp. 123–129. (Russian)

30. A. Yu. Borisovich, *The Plateau operator and bifurcation of two-dimensional minimal surfaces*, Global Analysis and Mathematical Physics, Voronezh Univ. Press, Voronezh, 1987, pp. 142–154. (Russian)

31. W. Jaco and J. H. Rubinstein, *PL minimal surfaces in* 3-*manifolds*, J. Differential Geom. **27** (1988), 493–524.

32. ——, *PL equivariant surgery and invariant decomposition of* 3-*manifolds*, Adv. in Math. **73** (1989), 149–151.

33. A. O. Ivanov and A. A. Tuzhilin, *On the deformations of a manifold that lowers the volume with maximum speed*, Vestnik Moscov. Univ. Ser.

Mat. **1989**, no. 3, 14–18; English transl. in Moscow Univ. Math. Bull. **44** (1989).

34. I. V. Shklyanko, *The one-dimensional Plateau problem on surfaces*, Vestnik Moskov. Univ. Ser. Mat. **1989**, no. 3, 8–11; English transl. in Moscow Univ. Math. Bull. **44** (1989).

35. R. G. Brown, *Enumeration of triangulations of the disk*, Proc. London Math. Soc. (3) **14** (1964), 746–768.

36. F. Harary and E. M. Palmer, *Graphical enumeration*, Academic Press, New York-London, 1973.

37. L. S. Polak (ed.), *Variational principles of mechanics*, Gosudarstv. Izdat. Fiz.-Mat. Lit., Moscow, 1959. (Russian)

38. V. I. Arnol′d, *Mathematical methods of classical mechanics*, "Nauka", Moscow, 1989. (Russian)

39. J. Eells and L. Lemaire, *A report on harmonic maps*, Bull. London Math. Soc. **10** (1978), 1–68.

40. ——, *Selected topics in harmonic maps*, CBMS Regional Conf. Ser. in Math., vol. 50, Amer. Math. Soc., Providence, RI, 1983.

41. A. O. Ivanov and A. A. Tuzhilin, *Solution of the Steiner problem for convex boundaries*, Uspekhi Mat. Nauk **45** (1990), no. 2 (272), 207–208; English transl., Russian Math. Surveys **45** (1990), no. 2, 214–215.

42. Z. A. Melzak, *On the problem of Steiner*, Canad. Math. Bull. **4** (1961), 143–148.

Subject Index

Errata

(−9 means line 9 down from the top, and +4 means line 4 up from the bottom; not including chapter headings or footnotes)

Page	Line	Currently	Should Be
vii	-9	Nitsche ref.	add , rev. English transl. of Chaps. I–V, *Lectures on minimal surfaces.* Vol. I, Cambridge Univ. Press, Cambridge, 1989.
vii	+4	Osserman ref.	add (Dover)
1	-9	a soap is left...	a soap film is left...
3	+2	containing to $N(P)$	containing $N(P)$
9	-19	It is well known that critical points...	Critical points...
10	+6	Dougals	Douglas
10	+9	Jesse Douglas	Douglas
13	+1	We note that a system...	A system...
14	+1	is a networks on the sphere S^2	is a network on the sphere S^2
30	-1	original contour and,	original contour and which,
34	+1	$l_0, l_{1/2}n, \ldots, l_1$	$l_0, l_{1/2n}, \ldots, l_1$
49	+1	large surface	larger surface
49	+3	large sur-	larger sur-
49	+5	large connected minimal surface	larger connected minimal surface
50	-2	large connected minimal surface	larger connected minimal surface
54	-7,8	reduces to the solution of differential equation [8].	reduces to the solution of a differential equation [8].
55	-6	a differential inverse	a differentiable inverse
71	-3	is increased by a factor	is expanded* by a factor (footnote added: * "Increased" is poor; better, "expanded", "multiplied" or "changed".)

Page	Line	Currently	Should Be
75	-8,9	Riemann manifold	Riemannian manifold
83	-6	genus h	genus h
90	-5	Continuting this construction	Continuing this construction
90	-10,11	minimal surfaces	minimal surface
101	-7	complex plane c	complex plane $\mathbb{C}$
101	+18	P-function	$\mathcal{P}$-function
101	+15	$P(z) =$	$\mathcal{P}(z) =$
101	+8	P-function becausec	$\mathcal{P}$-function because
101	+5	$a = 2\sqrt{2\pi}P(1/2)$	$a = 2\sqrt{2\pi}\mathcal{P}(1/2)$
101	+4	if P' denotes	if $\mathcal{P}'$ denotes
101	+3	P-function P	$\mathcal{P}$-function $\mathcal{P}$
101	+2	$(Pdz, a/P')$	$(\mathcal{P}dz, a/\mathcal{P}')$
102	-2	$(Pdz, c/P')$	$(\mathcal{P}dz, c/\mathcal{P}')$
102	-7	P-function	$\mathcal{P}$-function
103	+1	diffeomorphism with an image	diffeomorphism onto its image
105	+9	smooth funciton	smooth function
137		1. Dubrovin ref.	...1986. English transl. Springer-Verlag, Part 1, GTM **93**, 1984; Part 2, GTM **104**, 1985; Part 3, GTM **124**, 1990.
137		9. Osserman ref.	...55–136; English transl., *A survey of minimal surfaces*, Van Nostrand Reinhold, New-York, 1969.
139		38. Arnol'd ref.	...1989; English transl. of 1st ed. Springer-Verlag, Berlin and New York, 1978; 2nd ed., 1989.
141	1st col., -19	Convex Hull, 60	Convex hull, 60
141	2nd col., +5	classification, 82	classification (of 2-dimensional), 82
142	2nd col., +3	P-function	$\mathcal{P}$-function